Bird[illegible]

C[illegible]

Dedicated to the
Conservation Programmes of BirdLife Cyprus and KUŞKOR

BirdLife Cyprus

HELM FIELD GUIDES

Birds of CYPRUS

Colin Richardson and Richard Porter

Illustrated by
John Gale, Mike Langman and Brian Small

HELM
LONDON • OXFORD • NEW YORK • NEW DELHI • SYDNEY

HELM
Bloomsbury Publishing Plc
50 Bedford Square, London, WC1B 3DP, UK

First published in Great Britain 2020

A catalogue record for this book is available from the British Library

Library of Congress Cataloguing-in-Publication data has been applied for

ISBN: PB: 978-1-4729-6084-9; ePDF: 978-1-4729-6086-3; ePub: 978-1-4729-6085-6

2 4 6 8 10 9 7 5 3 1

Maps by Julian Baker
Design by Julie Dando, Fluke Art
Printed and bound in India by Replika Press Pvt. Ltd.

Cover artwork
Front: Cyprus Warbler (Brian Small)
Back (top to bottom): Cyprus Scops Owl (Brian Small), Masked Shrike (Brian Small), Eleonora's Falcon (John Gale), Cyprus Coal Tit (Brian Small)

MIX
Paper from responsible sources
FSC® C016779

CONTENTS

FOREWORD

I am deeply honoured to have been asked to contribute the foreword to *Birds of Cyprus*, a field guide rooted in the tried and tested tradition of Helm Field Guides. There is no doubt that field guides of this quality and detail, featuring quite outstanding illustrations, are fundamental to encouraging an interest in birds, in their study and, ultimately, in their conservation.

The island of Cyprus, lying on the Black Sea/East Mediterranean Flyway, is of great strategic importance to the migrant breeding birds of Europe. It is also a centre of bird endemism, one of six global Endemic Bird Areas in Europe. Bird habitats on Cyprus range from coastal cliffs and scrub, to lowland farmland, wetlands, scrub and vineyards in the foothills, and to the substantial forests of the Troodos and Kyrenia (Girne) Mountains. As a basis for a conservation strategy, 34 Important Bird Areas (IBAs) have been recognised.

In conservation terms, Cyprus has received widespread negative publicity in relation to levels of illegal hunting and trapping, in particular of migratory passerines to supply what is still a substantial restaurant market. There have been some welcome recent successes in reducing levels of trapping, with trappers arrested and prosecuted. Local Non Governmental Organisations (NGOs) and others, with support from the BirdLife partnership, have been fundamental to this and in the institution of longer term measures for the protection of biodiversity. These include the establishment of the Natura 2000 network of protected sites and targeted management projects at key sites such as the Oroklini and Akrotiri marshes, as well as the Kleides Islands off the Karpasia Peninsula. Other initiatives include a collaborative project aiming to reduce the use of dangerous rodenticides, by encouraging breeding of Barn Owls as biological control agents. Accidental raptor poisoning remains a serious issue and plans are being developed for a combined anti-poisoning and vulture reintroduction programme.

In public awareness terms, extensive publicity has supported the conservation message and interventions with the government focused on the implementation of treaty obligations under the EU Birds and Habitats directives. The major problem, as in most Mediterranean coastal regions, involves over-exploitation of natural habitats for tourism, and this is where NGOs in Cyprus are concentrating their efforts. The recent signing of a Memorandum of Cooperation between BirdLife Cyprus and KUŞKOR, the Turkish Cypriot bird protection society, is a welcome step for better nature protection island-wide.

Finally, let me congratulate the authors, publishers and all those duly recognised in the Acknowledgements for their dedication in producing this guide. *Birds of Cyprus* will serve to encourage a love and respect by Cypriots of their natural world as well as support local and visiting birdwatchers, a growing and valuable economic addition to the island's out-of-season tourism industry.

Tasso Leventis

Cyprus Scops Owl *Otus cyprius* at roost near Pegeia, Paphos (Albert Stöcker).

ACKNOWLEDGEMENTS

In preparing this field guide we have received help and encouragement from a great many people. First and foremost our thanks go to our friend, Simon Aspinall, co-author of the second edition of *Birds of the Middle East*. We are especially grateful to Mark Easterbrook, Chris Holt, John Sanders and Bill Stacey who have supported us over the years and given us incentive to write this guide. We would particularly like to thank the following who made invaluable contributions and supplied records, some over many years (with sincere apologies to anyone inadvertently omitted): Neil Alderson, Pete Alderson, Mark Askew, the late Simon Aspinall, Liz Atkinson, Ray Atkinson, Ian Barton, Mike and Sue Blair, Ian and Kate Bloomer, Di and Andy Brannon, Mike Bulpitt, Steve Cale, Melis Charalambides, Stavros Christodoulides, Yiannis Christofides, Loukas Chrystophorou, Sue and Alan Colquhoun, Nigel Cottle, Alan and Lynn Crane, Paul Curry, Judy Dawes, Lance Degnan, Joe Donaldson, Tim Drew, Terry Duncan, Pat Dwyer, John East, Mark Easterbrook, Pete Evans, Peter Flint, Trevor Fordham, Annika Forsten, Frank Foster, Ian Frith, Filippos Georgiades, Jeff Gordon, Michael Gore, David and Orcilla Hadrick, Mike Hamzij, Markus Handschuh, Martin Hellicar, Peter Hellyer, Angus and Mary Hogg, Brian and Pam Holgarth, Johannes Honold, Raija and Keith Howard, Savvas Iezekiel, Stuart Irving, Kostas Kailis, Nikos Kassinis, Chris Kehoe, Andreas Kephalas, Alex Kirschel, Marios Konis, George Konstantinou, Chris and Denise Lamsdell, John Lovatt, Christodoulos Makris, Roy Mason, Frank Mogford, Cliff Morrison, Ian Murray, June Neal, Dave and Eileen Nye, Aaron Ofner, Jonathan Osborne, Ernie Parkes, Nick Pegler, Marios Philippos, Martin Pitt, Karl Proctor, Doug Radford, Tony Reeves, the late George Reszeter, Glyn Roberts, Kathleen Rosewarne, Tobias Roth, Phil Saunders, Mick Sharp, Bob Shiret, Richard Smart, Matt Smith, Kevin Spearing, George Spiridakis, Bill Stacey, Harry Stanford, John and Anne Stapley, Chris Stavrou, Patricia Stavrou, Albert Stöcker, Jane Stylianou, Victor Tjernberg, Alan Turtle, Alan Tye, Dave and Jan Walker, Andy Waller, Mike and Annie West, David Whaley, Alan Whitehead, Marny Willis, Jim and Christine Willson, Pete Wragg, Dave Yarranton, Dick Yates, plus hundreds of other friends, colleagues and correspondents who provided their records over the years.

We are very grateful for the help of Peter Flint who made many helpful comments and whose expertise on Cyprus birds has ensured the accuracy of the texts and maps, especially of the endemics and other breeding specialties. We are indebted to Peter Clement, Alan Dean, Paul Donald, Dick Forsman, Nikos Kassinis, Alison McArthur, Killian Mullarney and Yoav Perlman who helped to check the distribution and identification of birds that are distinctive to Cyprus. We are grateful to BirdLife Cyprus, particularly Martin Hellicar, Melis Charalambides, Jane Stylianou and Christina Ieronymidou, for their useful advice on the breeding maps, species' status and on the correct usage of Greek and Turkish place names; and to Robin Snape of KUŞKOR for his helpful comments on species' distribution and birdwatching sites in northern Cyprus. Both organisations kindly provided texts on conservation, and details on reporting poaching and on how to submit records.

We also thank Per Alström, Mike Blair, Lars Jonsson, Dick Newell, Roger Riddington, Andy Stoddart and Lars Svensson who provided comments on identification in *Birds of the Middle East*, which have been used extensively in this field guide. The three artists, John Gale, Mike Langman and Brian Small, are particularly thanked for their ornithological advice and consummate professionalism in the production of the plates, and for their tolerance and good humour in the face of last-minute additions and changes. In the quest for information and travels in the Middle East, over many years, BirdLife International, The Darwin Initiative, the Ornithological Society of the Middle East, the Caucasus and Central Asia, and the Wetland Trust have always given their support.

The authors would like to thank Jim Martin and Jenny Campbell at Bloomsbury and Julie Dando at Fluke Art for their tireless encouragement and support, Guy Kirwan and Craig Brelsford for their careful editing and proofreading of the text, Tasso Leventis for honouring us by writing the Foreword, and our wives Sylvie and Julia for supporting us throughout.

All that remains is to wish everyone safe travels and good birding.

Colin Richardson and Richard Porter

Male Cyprus Warbler *Sylvia melanothorax* at Cape Greco (Ian Bloomer).

INTRODUCTION

This field guide to the birds of Cyprus is an updated abridgement of the second edition of *Birds of the Middle East* (Porter and Aspinall 2010). Full coverage is given for all species known to have occurred in the wild, to mid 2019. Almost every year new species are added to the Cyprus checklist.

About 95 species of bird breed annually on Cyprus, this avifauna being mainly of Western Palearctic origin. Of the 405 species on the island checklist, more than 380 are migrants, of which more than 120 are vagrants. The definition of a vagrant varies between countries: for Cyprus, the BirdLife Cyprus Rarities Committee in general uses a minimum of ten records per decade as the 'cut-off' point above which a species ceases to be considered a vagrant, but as a rare visitor instead. This exceptional diversity is the result of the country lying on a migratory crossroads, with western and eastern flyways coming into contact with each other, and having no clear divide. Unsurprisingly, Cyprus is a popular destination for visiting birdwatchers, joining a small but growing number of nationals and resident expatriates also pursuing this most absorbing hobby. Hopefully this field guide will both enable and encourage many more people to enjoy identifying those birds that they encounter, whether in the garden, at salt lakes, on reservoirs, on the coast or in the mountains. Additional recommended reading dealing with Cyprus and its birds, and also with Middle Eastern birds in general, is provided under 'References and further reading' (pp. 234–235). More detailed information on the status of the birds on Cyprus can be found in *Birds of Cyprus* (Flint and Stewart 1983, 1992).

The best birding sites are included in the section 'Birdwatching sites' (pp. 18–43) which covers all the major sites and how to find them.

We have prepared detailed distribution maps for all species (except vagrants) to a larger scale than previous books in the series, so that one can see at a glance the range of each species. The maps show whether it is a resident or migrant, and the approximate ranges of both passage migrants and winter visitors.

Our knowledge is constantly evolving, so if readers spot inaccuracies in the maps and texts, or require any further information, please write to richar@cytanet.com.cy.

TAXONOMY, NOMENCLATURE AND SEQUENCE

This is a very confusing and contentious area of ornithology, and newcomers to birding will have every reason to be mystified by the various treatments in different books. There are those taxonomists who prefer to 'lump' species and those who prefer to 'split' them (thus subspecies become species in their own right).

Species taxonomy and nomenclature (English and scientific names) follow the so-called IOC World Bird List (Gill and Donsker 2018) which is adopted by the Ornithological Society of the Middle East, the Caucasus and Central Asia for its 'OSME Region List' (ORL). The checklist, taxonomy and nomenclature differ slightly from the official Cyprus list produced by BirdLife Cyprus, which generally follows that published by HBW/BirdLife International. The IOC list is updated four times per annum and can be viewed online at worldbirdnames.org while the OSME list can be seen at osme.org/orl. For presentational reasons, the sequence of families and species does not follow the Gill and Donsker list precisely. See the Checklist on pp. 236–246 for the latest official order of species.

Only rarely have English names deviated from those proposed by Gill and Donsker, in each case involving logical divergence from those proposed by them, rather than (and despite temptation) any personal preference. We have also given alternative names where relevant (under 'Alt' at the end of the species text), but only those that are still frequently used or help to prevent confusion. Treatment of alternative names is not comprehensive but the scientific name should prevent ambiguity, although many of these have also been subject to change in recent years.

In vogue with the times, though not without substantial scientific backing, the taxonomic trend has been towards splitting species and the recognition of new species. With such a constant state of flux, it is impossible to stay ahead of developments. This field guide is not a taxonomic authority and has simply embraced the latest thinking on what may or may not constitute a species. However, we describe all recognisable taxa known to occur on Cyprus.

ILLUSTRATIONS AND IDENTIFICATION TEXT

Our aim has been to make the illustrations and species accounts of use to beginner and expert alike. We have concentrated on those features important for identification and, accordingly, these are italicised in the text. For each species, the length from bill tip to tail tip (L) is given in centimetres; for larger birds wingspan (W) is also given. Where the identification of a species does not present a problem, the texts are often brief, whereas more difficult species have necessarily warranted more detailed descriptions, sometimes including biometrics.

There is no need to stress the increasing importance of digital photography in helping to make correct identifications, especially of tricky species. This enables examination and discussion with experts at a later date.

Abbreviations used on the plates: ad – adult, juv – juvenile, imm – immature, ♂ – male, ♀ – female.

BIRD TOPOGRAPHY

The illustrations below show the various features of a bird – its topography – used in the identification texts. Knowledge of these, especially the feather tracts, is vital for describing a bird and its plumage. Making field sketches, even if embarrassingly poor, is a good way to learn.

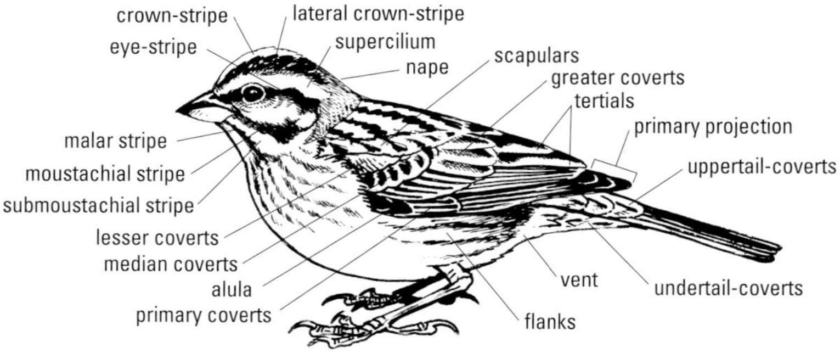

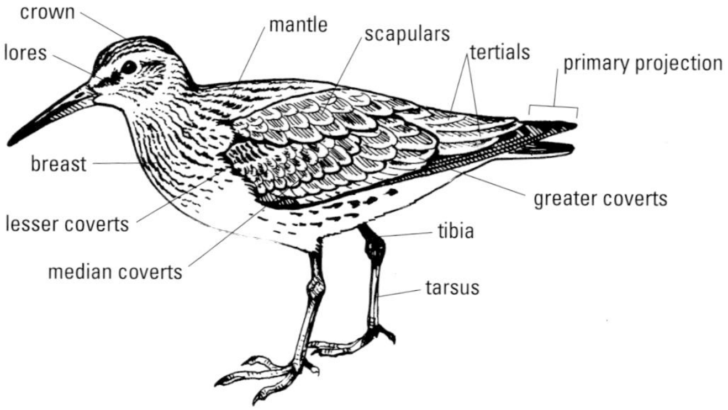

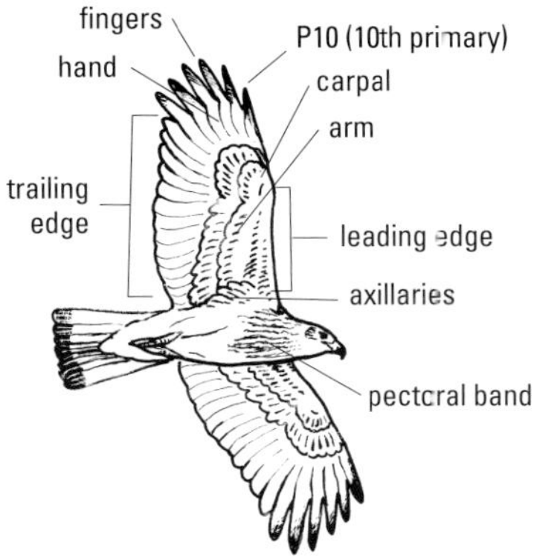

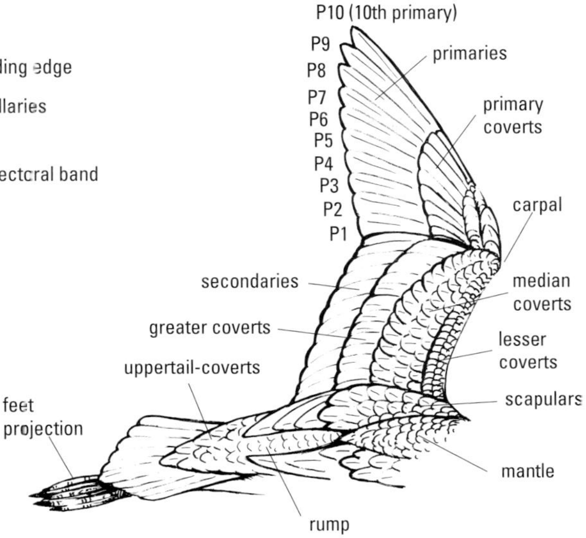

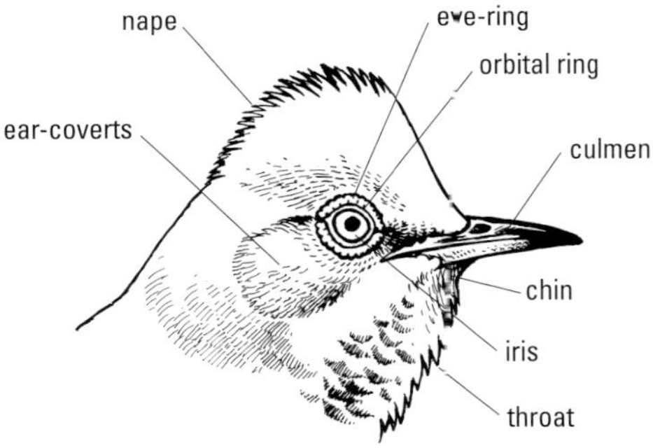

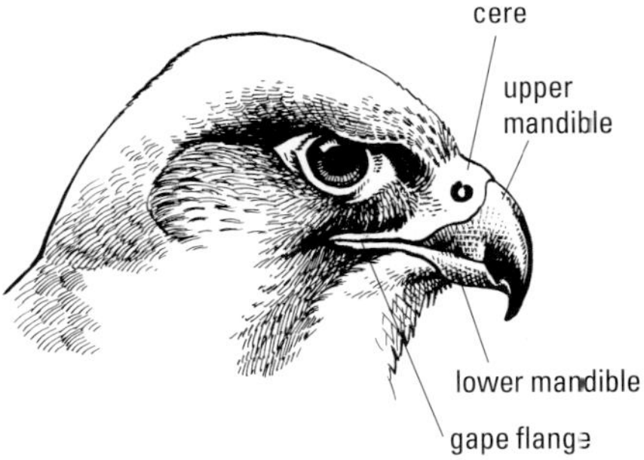

VOICE

Calls and songs are given where a species occurs regularly in the region or where these might aid identification. Song is, of course, described for all species breeding on Cyprus, but many migrants also give at least occasional bursts of song or sub-song on passage or in winter, despite breeding elsewhere. Where this is known to be the case on Cyprus then a description of those songs is also given.

The internet can be checked for sound recordings. There is no substitute for learning vocalisations or, if unable to commit them to memory, in comparing recordings. However, the use of technology to record calls and songs has somewhat lagged behind digital photography as a means for confirming the identity of a bird. Few of us use sonograms, but the way forward certainly lies in this field. In our region, for example, the calls of the different subspecies of Common Chiffchaff may only be separable by such analysis. Indeed, many recently described species of birds (and even some mammals) have first been noticed by their different vocalisations.

There is also, as amply demonstrated by our own species, *Homo sapiens*, the matter of regional dialects. Songs and calls of bird species commonly found in Western Europe may often sound quite different in the Middle East. Even within the region, strong intraspecific variations can frequently be noticed.

STATUS, DISTRIBUTION AND HABITAT

At the start of each account we have summarised the species' status on Cyprus, i.e. how common it is and when and where it is likely to occur. This will immediately inform the reader how likely or not an observation might be, especially the chance of the species occurring in a certain habitat or at a certain time of year. You should remember that most birds you see are likely to be common, but if it is a rare species, or at an odd time of the year then it will be worthy of more scrutiny, and perhaps worth notifying local birders or writing a description for the country's bird recorder.

Upper and lower case letters are used to indicate relative abundance of a species; upper case reflecting larger numbers or greater abundance either at that season and/or in comparison with other closely related species. When PM and WV (or pm and wv) share the same case, the season when that species is commoner appears first, unless they are actually equally common at both seasons:

RB/rb – resident breeder
MB/mb – migrant breeder
ob – occasional breeder
FB – former breeder
PM/pm – passage migrant
WV/wv – winter visitor (Dec–Feb at least)
v – vagrant (fewer than ten records per decade)
E/I – escapee or introduction

The habitats described in the text are those occupied on Cyprus and this may help in species identification, or at least narrow down the list of possible contenders, and direct you where to look for a particular species. We have described the likely habitat for most species. Some vagrants, by their very nature, are off-course and often found on coasts and headlands, in unsuitable habitat or, in the case of flying birds of prey, almost anywhere, so no habitat section has been included in the text.

We have elected not to include a handful of species that are on the official Cyprus list, as we feel the evidence to support their acceptance is lacking. These are Desert Lark *Ammomanes deserti*, Dupont's Lark *Chersophilus duponti*, Graceful Prinia *Prinia gracilis* and Striolated Bunting *Emberiza striolata*. To keep up to date with the official Cyprus list and status codes, readers are directed to birdlifecyprus.org/publications.

MAPS

The distribution maps distinguish between resident and migrant breeders and also show the passage and/or winter range for all regularly occurring species. Vagrants are not mapped. Note that the maps portray current known ranges (not historical ones). The breeding, passage and winter distributions of birds in Cyprus are still being studied and many gaps exist, shown on the maps by a question mark '?'. For some common species the colour tone covers an exaggerated range due to the patchwork nature of the island's habitats, but obviously a species is most likely to occur in suitable habitat, within the range shown. In the eastern Mediterranean where the seasons are not clear-cut, autumn migration may continue until Dec, while spring migrants may reappear as early as Jan. In general, however, a species shown by hatching is generally a passage migrant that is absent in winter or present then only in relatively small numbers.

Note that many non-breeding waterbirds, often immatures, may be found year-round at wetlands, including in particular many different shorebirds, among them even Arctic-breeding species. Do not be surprised to see, for example, Ruff or Little Stint at salt lakes in Cyprus in midsummer.

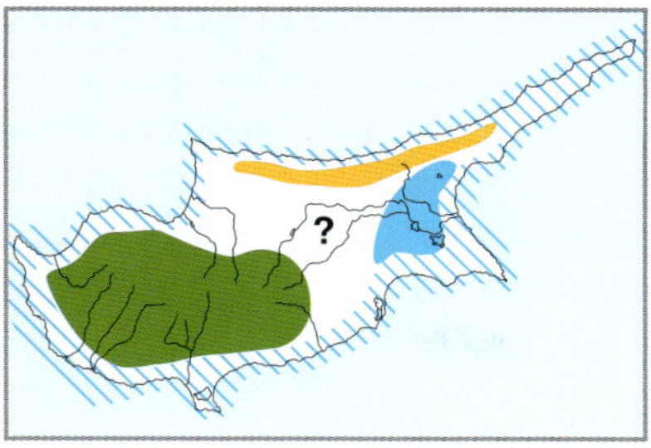

Green – resident (present all year, may also occur as a breeding visitor)

Blue – winter visitor (may also occur as a passage migrant)

Orange – breeding visitor (or occasional breeder)

Blue hatch – passage, or local wanderer when not breeding

? – may breed

IMPORTANT DAMS AND RESERVOIRS

Cyprus has an intense Mediterranean climate with the seasonal rhythm strongly marked in respect of temperature, rainfall and weather generally. There are hot, dry summers from mid May to mid Sep and rainy, rather variable winters from Nov to mid Mar (source: Government of Cyprus, Dept. of Meteorology). As a result, many important wetlands dry out completely, sometimes for years, and this affects the occurrence and abundance of waterbirds. Generally, salt lakes are seasonal and are likely to be wet only after winter rainfall. The locations of the most interesting and productive dams and salt lakes are shown on the map.

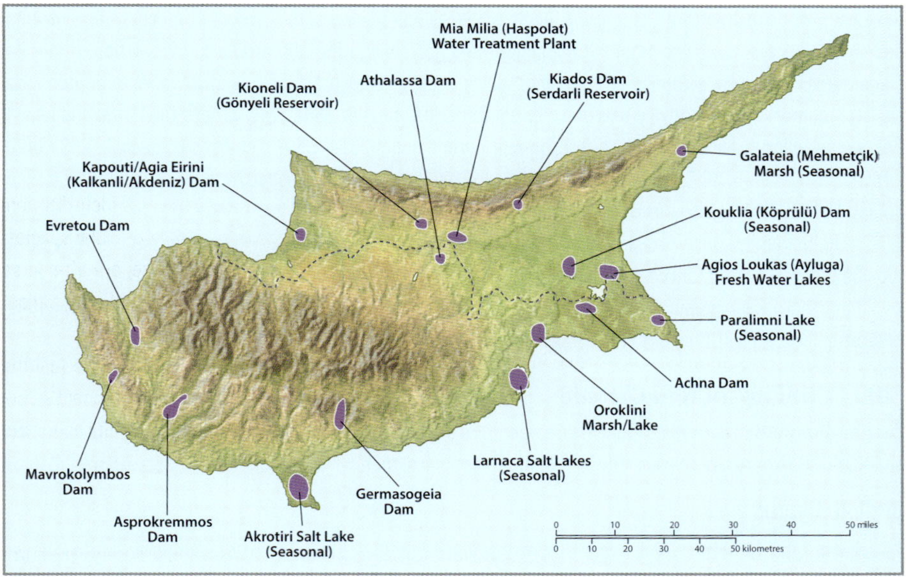

Map 1: Important dams and reservoirs in Cyprus.

CODE OF CONDUCT AND SUBMITTING RECORDS

The welfare of birds must come first, especially of nesting birds. Try to minimise time spent near nests or chicks; even if you haven't actually located them, it is usually evident from the behaviour of the adult(s) that are nearby. Photographers please take note! Roosting and feeding birds, notably waterbirds, are easily disturbed so please be thoughtful when, for example, trying to get that frame-filling shot of a Greater Flamingo.

Bird records are invaluable for Cyprus and help build a more complete picture of the bird fauna of the island, thus making a key contribution to ongoing conservation efforts. Whether you are a resident or a visitor, birdwatching regularly or just enjoying the birds you see when out walking, keeping records of what you see is important. Please submit your sightings to the BirdLife Cyprus Bird Recorder (birdlifecyprus.org/record-your-sightings) or through the Cyprus portal of BirdTrack (blc.birdtrack.net). Submitting records monthly to the Recorder will enable your records to be added to the country database. Extra details may be required of rare birds and a list of those species can be found on birdlifecyprus.org. In northern Cyprus, bird records should also be sent to KUŞKOR at info@kuskor.org. See kuskor.org for more details.

BIRD KILLING IN CYPRUS

Legal hunting is permitted on specific Wednesdays and Sundays (although some areas allow daily shooting) in set areas from Sep to Feb, with a crow cull in early summer. Despite the appeals of conservation groups, Turtle Dove remains on the target list in early autumn. In late autumn Black Francolin, Woodcock, Chukar and hare are targeted, until after New Year when thrushes are taken. On permitted hunting days exclusion zones around wetlands are often disregarded, birds of prey are regularly found shot and Bonelli's Eagles are among those reported at least annually. Poison against stray dogs and foxes also impacts birds of prey and rodenticides affect many owls.

Poaching and trapping is a major problem on Cyprus. BirdLife Cyprus has monitored and campaigned on this issue for many years, with some success. Due to a relative lack of capacity among conservation groups, less has been done to monitor and tackle illegal trapping in northern Cyprus, although with the support of the Committee Against Bird Slaughter (CABS), KUŞKOR has gained some ground in recent years, making dedicated surveys, removing trapping equipment and galvanising the authorities to take action. If you do come across unsightly evidence of poaching, which occurs throughout the island to variable degrees, do NOT try to intervene. We advise that you contact BirdLife Cyprus at +357 2245 5072 or the Game and Fauna Service as follows: Nicosia: 22662428, 22664606 (office hours), 99445697; Lemesos (Limassol): 25343800 (office hours), 99445728; Larnaca/Famagusta: 24805128 (office hours), 99634325; Paphos: 26306211 (office hours), 99445679. It is also worth calling your local police station.

In the Sovereign Base Areas (SBAs), the responsible authority is the SBA Police: for the Dhekelia SBA, the number is 1443 and for the Akrotiri SBA, the number is 25967262.

In northern Cyprus you should contact KUŞKOR on +90 542 881 3750 (info@kuskor.org) or the hunters federation police AVKOR on direct line 140.

BIRD CONSERVATION IN CYPRUS

There are two bird conservation NGOs on Cyprus – BirdLife Cyprus, the official partner of BirdLife International, and KUŞKOR in northern Cyprus. These two organisations signed a Memorandum of Cooperation in 2018.

BirdLife Cyprus

Founded in 2003, BirdLife Cyprus works to conserve wild birds, their habitats and wider biodiversity in Cyprus, through research, monitoring, lobbying, and conservation and awareness-raising actions. It is very active in running campaigns against illegal bird trapping and poaching, and for the designation and protection of Important Bird Areas as Special Protection Areas, as well as campaigns in the areas of agriculture, education and awareness-raising.

Cyprus faces growing conservation challenges on several fronts: loss, fragmentation and degradation of habitats due to badly planned developments and lack of sound site management, illegal bird killing and high hunting pressure, climate change and invasive alien species, especially acacias. All of these are very real threats to the island's birds and other wildlife. BirdLife Cyprus (*Πτηνολογικός Σύνδεσμος Κύπρου*), the partner organisation of BirdLife International, has been working on several fronts to try to achieve better protection and conservation of the island's precious bird fauna. While there have been notable successes in securing protected status for key sites and in tackling illegal bird trapping, many challenges remain.

Protecting key habitats is the cornerstone of the BirdLife Cyprus effort. All Important Bird Areas (IBAs) identified by the organisation now enjoy legal protected status as Natura 2000 sites, at least in the Republic. This is a great success, but not one that always translates into effective protection from damaging road, housing, tourism or other developments, despite BirdLife Cyprus's constant lobbying. Indeed, Natura 2000 sites such as the island's two salt lakes, the Troodos and Paphos forests, Cape Greco and Akamas, do not as yet even enjoy protective management. BirdLife Cyprus has tried to lead by example in the field of site management, coordinating successful projects for the restoration of both Oroklini Marsh and Akrotiri Marsh in recent years (both key IBAs).

Tackling the persistent problem of illegal bird trapping is the organisation's most high-profile campaign This long-running effort has led to a significant reduction in the damaging use of mist-nets and lime-sticks. Trappers target mainly Blackcaps and Song Thrushes, for the lucrative illegal market in *ambelopoulia* 'delicacies', but the indiscriminate nature of trapping means more than 150 migrant and local species are known to be caught and killed. Generating the political will to tackle the law-breaking tavernas still offering *ambelopoulia* continues to prove difficult. But there have been notable recent enforcement successes in the Eastern Sovereign British Base (SBA) of Dhekelia, once a notable trapping black-spot. The Royal Society for the Protection of Birds (RSPB, BirdLife in the UK) investigations team has played a key part in an effective collaborative effort between BirdLife and the SBA authorities that has led to serious prosecutions. The BirdLife Cyprus anti-trapping programme includes systematic monitoring of trapping activity, working with enforcement authorities in the SBAs and Cyprus Republic, lobbying in Nicosia, London and Brussels, and education and awareness-raising focusing on local schools.

Many in Cyprus still view trapping as a 'harmless tradition', a dangerous misconception that BirdLife Cyprus is working to change. Indeed, education and awareness-raising actions have come to play an increasingly large part of the work of BirdLife Cyprus, with the focus on reconnecting people – and children in particular – to nature.

The Tasso Leventis Conservation Foundation has been the cornerstone of support for BirdLife Cyprus's work, while support from the BirdLife partners in the UK (RSPB) and Germany (NABU) has been key to what has been achieved in education and in the fight against bird trapping. BirdLife Cyprus has a memorandum of cooperation with Turkish Cypriot bird protection society KUŞKOR, which enables joint action for conservation across the island. KUŞKOR and BirdLife Cyprus have already collaborated on key projects, including the production of the latest IBA inventory for Cyprus, published in 2014.

Last but not least, BirdLife Cyprus has established monitoring programmes to keep track of bird populations. These include a common breeding bird programme, monthly counts at all key wetlands, nocturnal species and migrant raptor surveys, as well as monitoring for key species such as Griffon Vulture, Audouin's Gull and European Roller. Birdwatching is also a key activity for BirdLife Cyprus, harking back to its roots as a society. Field trips for members and friends are organised on a weekly basis; for further information see birdlifecyprus.org

KUŞKOR

Despite their long-term isolation from international conventions on nature protection, significant efforts have been made by Turkish Cypriots to declare and manage important areas for conservation. The first dedicated surveys of seabird nesting sites were carried out by the Near East University and Cyprus Ecological Society, and continue through KUŞKOR (the Society for the Protection of Birds and Nature) and BirdLife Cyprus. Using data generated through the work of KUŞKOR and other NGOs, an EU-funded project led to the designation of seven potential Natura 2000 sites in the late 2000s. Although these will not be ratified until the EU community acquis applies, the northern Cyprus Environmental Protection Department and Town Planning Departments have incorporated the Natura 2000 area management plans into current legislative documents. In 2014, six IBAs were declared by BirdLife International through data provided by

KUŞKOR and BirdLife Cyprus. The Animal Husbandry Department, which manages fisheries, has supported KUŞKOR by banning landings at the marine IBA seabird nesting islands and set fisheries legislation to protect sea turtles and monk seals. KUŞKOR and other environmental NGOs lobby collaboratively via an Environment Platform Group.

KUŞKOR has been active since 1988, and strives to protect birds and their habitats in northern Cyprus through education, research, monitoring and lobbying, sharing similar objectives and challenges to Birdlife Cyprus. KUŞKOR runs an office and Nature Library in Geunyeli (Gönyeli) and has an active committee of volunteers. While KUŞKOR has no permanent paid staff, it sustains capacity via grants and donations. It has an active membership, offers free classes to schools, undertakes research and monitoring, organises field trips and expeditions for BTO ringing teams, undertakes monthly field trips and moderates a bird sightings forum (Yahoo group), which can be accessed via their website. They also man a stand at the UK Birdfair every August to promote ecotourism in northern Cyprus. Sightings are collated to inform conservation planning, and exceptional or important records are published in the journal *Sandgrouse*.

BIRDWATCHING SITES

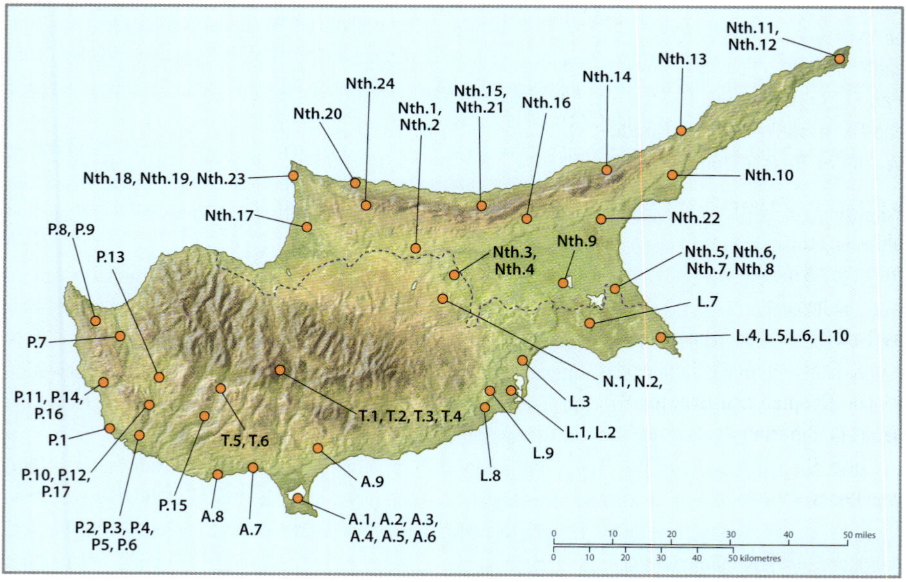

Map 2: Key birdwatching sites in Cyprus.

Paphos District

P.1. Paphos Lighthouse 34°45'21.9"N 32°24'29.4"E (34.756080, 32.408153) Best mid Feb–May inside the archaeological site; Jul–Feb on the headland outside the fence. Located near the Paphos tourist area, official entry is from Paphos Marina. Entrance fee €4.50, open 08.30–17.00h (16 Sep–15 Apr); 08.30–19.30h (16 Apr–15 Sep). In spring, the tree-lined avenues and bushy areas host a large selection of migrants, especially Eurasian Hoopoe, Eurasian Wryneck, redstarts, nightingales, thrushes, warblers, flycatchers and shrikes, while the open areas, including the archaeological sites, attract larks, wagtails, pipits, Siberian Stonechat, wheatears and buntings. Tired migrants such as Short-eared Owl and Common Quail arrive over the sea and in late spring Black-crowned Night and Purple Herons roost overnight in the tallest trees. From late summer to early spring, a walk around the headland (outside the fence) should provide Greater

Paphos Lighthouse and its headland are regular sites in spring for shrikes, wheatears, larks, warblers and wagtails, and in autumn and winter for Greater Sand Plover (Colin Richardson).

Sand Plover and other waders. The sea off Cyprus is generally rather quiet and most gulls are likely to be **Yellow-legged**, but occasionally **Audouin's** and **Pallas's Gulls** pass; virtually the only tern likely in winter is **Sandwich Tern**. Recent rarities have included: **Little Bustard**, **Cream-coloured Courser**, **Pacific Golden Plover**, **Black-legged Kittiwake**, **Caspian Tern**, **Blue-cheeked Bee-eater**, **Richard's** and **Olive-backed Pipits**, **Hooded Wheatear**, **White-throated Robin**, **Asian Desert** and **Olive-tree Warblers**, **Rose-coloured Starling**, **Isabelline Shrike**, **Trumpeter Finch** and **Cinereous Bunting**.

P.2. Mandria 34°41′53.0″N 32°31′49.0″E (34.698045, 32.530274) Located approximately 15km south-east of Paphos town on the old Limassol road, 2km past the turn-off for Paphos airport. Mandria is well signed from the A6 motorway and the old Limassol road (B6).

The habitat is farmland, with mature trees, olive groves and ploughed fields. The promontory known as 'Lark Corner' has attracted seven species of lark. Rocky islets provide safety for local **European Shag** and coastal waterbirds. Spring passage in late Feb–May is the best time, when almost annual specialties include **Caspian** and **Greater Sand Plovers**, **Blue-cheeked Bee-eater**, **Bimaculated Lark**, **Citrine Wagtail**, **Siberian Stonechat** and **Desert Wheatear**. In autumn, Sep–mid Oct is a good time for raptors, but also **Collared Pratincole**, **Yellow Wagtail**, **Greater Short-toed Lark**, bee-eaters, hirundines, wheatears, shrikes and coasting herons and ducks; in Oct–late Feb major passage involves **Calandra Lark**,

The farms and coastline at Mandria are hotspots for migrants from late autumn to spring, with rarities occurring regularly (Colin Richardson).

The irrigated fields at the Paphos Sewage Plant are a magnet for migrants, and host several pairs of Spur-winged Lapwings (Colin Richardson).

Red-throated and **Richard's Pipits**, **Spanish Sparrow** and winter visitors from late Oct. It is easy to drive slowly around the farm tracks, but walking is likely to produce more birds. Recent rarities include **Cream-coloured Courser**, **Black-bellied Sandgrouse**, **Rufous Turtle Dove**, **Isabelline**, **Red-tailed** and **Steppe Grey Shrikes**, **Hooded Wheatear**, **'Masked' White Wagtail** and **Blyth's Pipit**.

P.3. **Paphos Sewage Plant Fields** 34°43'42.9"N 32°27'56.4"E (34.728595, 32.465665) Located 10km south-east of Paphos town. Take B6 (the airport road) from Paphos to the village of Acheleia, then at the bus stop (opposite the Akis taverna) turn towards the sea, following the asphalt road for 2km, before turning right at the airport boundary, to reach the sewage plant buildings after 600m. Interesting all year, but particularly Aug–May. Habitat is cereal and animal fodder fields, interspersed by lines of reeds, olive, almond and citrus groves, and bordered by mature tall cypress trees. The fields are irrigated by recycled water from the water treatment works, and the best time is immediately after a field has been cut, when pipits, wagtails and larks abound in Sep–Mar. In addition to the resident group of **Spur-winged Lapwings** (which are often inside the works) waders such as **Northern Lapwing**, **Ruff**, **Little Ringed Plover** and **Eurasian Curlew** are also likely. The nearby Ezousas River is interesting after spring rains and is often patrolled by birds of prey, which also perch on the power lines. There are reedbeds at the river mouth. Key species present all year include **Cattle Egret**, **Bonelli's Eagle**, **Black Francolin**, **Eurasian Stone-curlew** and **Zitting Cisticola**; in spring **European Bee-eater**, **Red-rumped Swallow**, **Red-throated** and **Tawny Pipits**, **Citrine Wagtail**, **Masked** and **Woodchat Shrikes**; in autumn **Pallid** and **Montagu's Harriers**, **Steppe Buzzard**, **Lesser Kestrel**, **Red-footed Falcon**, **Merlin**, **Bluethroat** and **Lesser Grey Shrike** occur. Recent rarities have included **White-throated Kingfisher**, **Isabelline Shrike**, **Blue-cheeked Bee-eater**, **Richard's Pipit**, **Siberian Stonechat**, **Dusky Warbler** and **Rustic Bunting**. *The fields and sewage plant share a border and fence with Paphos International Airport so beware using camera equipment, as the area is sometimes patrolled by security police.*

P.4. **Asprokremmos Dam and Foinikas** 34°43'39.8"N 32°33'08.2"E (34.727722, 32.552279) The dam is located c.16km east of Paphos town, approached from the old Limassol road (B6). The turn-off is 1.3km east of the Mandria exit (Eko petrol station); the access road is signposted. There is a car park on the left before the dam wall, set within mixed woodland. The dam at first seems rather barren, but in autumn and winter you can find distant waterfowl, birds of prey and gulls from its wall. Key species in the wood in spring

The deserted village of Foinikas, on the Asprokremmos Dam, hosts Finsch's Wheatear and Blue Rock Thrush in winter (Colin Richardson).

include **Great Spotted Cuckoo**, **Eurasian Hoopoe**, **Eurasian Wryneck**, **Eastern Bonelli's Warbler**, **Semi-collared Flycatcher** and **Golden Oriole**; on the dam **Great Crested** and **Black-necked Grebes**, **Garganey** plus other ducks and assorted gulls. From the weather station above the wood, it is possible to see **Long-legged Buzzard**, **Bonelli's Eagle**, **Lesser Kestrel**, **Red-footed Falcon** and **Alpine Swift**. The scrubby habitat as far as the Anarita Mast is good for wheatears, **Spectacled** and **Cyprus Warblers**, larks, pipits and buntings. In winter, many waterfowl, **Woodlark**, **Blue Rock Thrush** and **Goldcrest** are likely. In autumn, species are similar to spring, plus **European Nightjar** and **Lesser Grey Shrike**. Up to ten **Eleonora's Falcons** hunt here most days Jul–Oct. Rarities have included **Greater Scaup**, **Egyptian Goose**, **Pallas's Gull**, **White-breasted** and **Pied Kingfishers**, and **Rose-coloured Starling**. At the shallow end of the dam, the abandoned village of Phinikas 34°45'11.9"N 32°34'10.2"E (34.753299, 32.569501) is accessed from Nata along the dry riverbed via a rough track. There is also a rutted downhill track starting at the Asprokremmos Dam Water Treatment Plant (1.5km east of Anarita, and well signed). Halfway down, this track provides good views of any waterbirds at the shallow end. In winter Phinikas hosts several **Finsch's Wheatears** and **Blue Rock Thrush**. **Rock Sparrow** has also occurred.

When the dam overflows, wetlands are created downstream (Lower Xeros Potamos), which can be an exceptional draw for waterbirds in spring.

P.5. Agia Varvara 34°45'14.3"N 32°30'27.9"E (34.753967, 32.507756) Located 10km east of Paphos. From the A6 motorway, take the third turn-off from Paphos signposted 'Agia Varvara', follow signs to the village and head to the river ford. There are large pools (soakaways) either side of the river north of the ford which are filled and emptied regularly. Best Mar–May and Sep–Nov. A wide valley and seasonal river (usually flowing Dec–Apr) with reedbeds, tamarisks, tall grasses and bushes. Throughout the year **Bonelli's Eagle**, **Long-legged Buzzard** and **Black Francolin**. In spring the river and soakaways have **Little Bittern**, **Baillon's**, **Spotted** and **Little Crakes**, and **Great Snipe**, plus waders and herons. The valley is a good place to see **European Roller**, **Red-rumped Swallow**, **Golden Oriole** and **Ortolan** and **Cretzschmar's Buntings**. In winter, **Finsch's Wheatear** is present on the north-west hillside.

P.6. **Anarita Park** 34°45'46.8"N 32°32'13.7"E (34.763002, 32.537137) Located 12km east of Paphos. On the A6 motorway from Paphos take the third exit, signposted 'Agia Varvara, Amargeti, Axylou' etc. Head uphill on the main road (E606) for c.2.5km, past the turn-off for Agia Varvara, until you reach the second chicken factory sign, then bear right along the graded track following the electricity power lines. The whole area over the next few kilometres is known as 'Anarita Park'. Best late Oct–May. This area of short grassy foothills and boulders is a regular wintering site for **Finsch's Wheatear** and **Blue Rock Thrush**, and **Common Rock Thrush** is present in spring. It is a good site to see **Bonelli's Eagle** and **Long-legged Buzzard** all year, while **Lesser Kestrel** and **Red-footed Falcon** perch and hunt from the wires near the chicken factory, and **Hen, Montagu's** and **Pallid Harrier** pass through in spring and autumn. Other birds include **European Roller, Great Spotted Cuckoo, Alpine Swift, Eurasian Wryneck, Cyprus Wheatear, Cyprus Warbler, Red-throated Pipit** and **Cinereous Bunting**. Rarities have included **Levant Sparrowhawk, Amur Falcon, Asian Desert Warbler** and **Red-fronted Serin**.

The boulder-strewn hillside at Anarita Park hosts Lesser Kestrel, Red-footed Falcon, Finsch's Wheatear, and Blue and Common Rock Thrushes in season (Colin Richardson).

P.7. **Evretou Dam** 34°57'29.7"N 32°29'07.6"E (34.958262, 32.485451) To reach the shallow end of the dam, follow the B7 Paphos–Polis road (via Tsada) and turn right on the E712 'Simou & Lasa' at the 'Watermill' café approximately 3.5km north of Giolou. After 1km, turn left onto the rough concrete track uphill, pass the round water tank, bear right, and drive steeply down to the dam, at the abandoned village of Evretou. For a smoother, but longer route, continue on the E712 for a further 3.7km, turn left for 'Sarama' downhill to the valley bottom, and turn left again along a tree-lined river for 2.5km to the shallow end of the dam. Best Aug–May. In addition to waterfowl, waders and herons, **Bonelli's Eagle** and **Long-legged Buzzard** are likely year-round. **Finsch's Wheatear** and **Blue Rock Thrush** often winter near the deserted village (or on the dam wall at the north end) while **Eurasian Crag Martin** sometimes feeds over the dam late Oct–Mar. Upstream, the riverine alders and plain host **Eurasian Siskin**, while **Eurasian Woodcock** occasionally winters on the riverbank in the brambles and long grass. In early autumn **Cretzschmar's** and **Ortolan Buntings** feed on the shore. Surprise visitors in recent years have included **White Pelican, Pygmy Cormorant, Pied Kingfisher, Rufous-tailed Bush Robin** and **Trumpeter Finch**.

The shallow end of Evretou Dam and the deserted village of Evretou (Colin Richardson).

Secondary sites in the Paphos area

P.8. Neo Chorio–Smigies 35°01′23.0″N 32°20′50.2″E (35.023042, 32.347274) On the edge of the unspoilt Akamas Peninsula, the habitat is scattered carob, almond and olive groves, open cereal fields and meadows. Take the concrete road from Neo Chorio to Smigies picnic site (and nature trail), about 3km. Best late Mar–early May and Sep–Oct. Species in spring include **European Roller**, **Great Spotted Cuckoo**, **Masked Shrike** and **Cretzschmar's Bunting**. **Cyprus Scops Owl** nests in the woodland, and can be heard calling in the daytime. Most *Sylvia* warblers (except **Cyprus Warbler**) and *Ficedula* flycatchers are regular. Also **Eastern Bonelli's Warbler**, **Golden Oriole**, **Ortolan Bunting** and **Lesser Grey Shrike** in autumn. Birds of prey are regular including **Northern Goshawk**, **Lesser Kestrel**, **Red-footed** and **Eleonora's Falcons**.

P.9. Baths of Aphrodite 35°03′27.9″N 32°20′46.9″E (35.057744, 32.346367) A caravan park among mature trees, situated on the cliff edge overlooking Chrysochou Bay. Key species in spring include **Collared** and **Semi-collared Flycatchers** and **Eastern Bonelli's Warbler**. Other regulars include breeding **Cyprus Wheatear**, **Masked Shrike** and **Pallid Swift**, the latter nesting in the cliffs and on rocky islets. In Aug–Oct seawatching is recommended, with groups of **Garganey**, herons, egrets and **Glossy Ibis** arriving in Chrysochou Bay, before flying high over the peninsula. **Wallcreeper** has wintered on the cliffs.

P.10. Armou Hills 34°48′26.0″N 32°29′23.0″E (34.807222, 32.489722) At the Paphos end of the motorway, at the large roundabout, head uphill on the E710. After 5km there is a mini-market (Point Kiosk), continue 500m uphill taking the next left-hand turn to Armou; then turn right heading uphill on a narrow concrete road. Habitat is farm fields, deep valleys, gorse and scrub. Key birds include **Long-legged Buzzard**, **Bonelli's Eagle**, **Eurasian Hobby**, shrikes, wheatears, warblers, buntings, winter larks and thrushes. **Cyprus Scops Owl** can be heard calling in Armou village late Feb–Jun.

P.11. Avagas Gorge 34°55′14.0″N 32°20′16.1″E (34.920560, 32.337794) A steep-sided narrow gorge with a lush perennial stream. Park in the car park at the river crossing near the coast, or if in a 4×4, drive inland

The narrowest point of the Avagas Gorge, where Wallcreeper is regular in winter (Colin Richardson).

2km along the rough track (below Viklari taverna) to the car park at the start of the gorge. The bushes host thrushes and warblers including **Eastern Bonelli's**, but in Nov–Feb the main target is **Wallcreeper**, which feeds on the steep cliffs. Finding one can involve clambering quite far into the narrowest part of the gorge. **Eurasian Wren** breeds in the gorge.

P.12. Episkopi and Ford 34°47'50.7"N 32°31'47.1"E (34.797425, 32.529735) From the Paphos motorway end take the turn-off for Episkopi, travel uphill past Armou, then down into the Ezousas Valley (11km). When flowing, the river passes under a concrete ford, flanked by overgrown grasses and reeds; a good site for herons, crakes and **Moustached Warbler**. **Peregrine Falcon** and **Blue Rock Thrush** are often found on the dramatic cliff in the village. A drive upstream for 2km along the surfaced road to the forested picnic area and iron bridge can be interesting. **Bonelli's Eagle** often soars over the hills to the north.

P.13. Kannaviou Dam and Forest 34°56'09.0"N 32°34'32.2"E (34.935826, 32.575606) This dam on the very edge of the Paphos Forest hosts few waterfowl, but from the wall there are chances of **Northern Goshawk**, **Eurasian Crag Martin**, **Cyprus Wheatear** and **Black-headed Bunting** in late spring. Take the graded forest track signposted 'Sarama' – the turn-off is 500m south of the dam wall towards Kannaviou village – and travel 1–2km into the forest to look for **Cretzschmar's Bunting** and the local subspecies **Cyprus Coal Tit**, **Cyprus Jay** and **Cyprus Wren**, all of which are present here.

P.14. Kathikas (vineyards) 34°53'59.1"N 32°26'25.3"E (34.899756, 32.440362) This area of vineyards and farms lies within a triangle bounded by the Kathikas–Akoursos–Stroumbi roads. At an altitude of 650m, it holds a high density of breeding **Black-headed Bunting**, in song mid Apr–June. **Long-legged Buzzard** and **Bonelli's Eagle** often fly over, and other key breeding species are present, mainly in spring, including **Black Francolin**, **European Roller**, **Great Spotted Cuckoo**, **Cyprus Wheatear** and **Cyprus Warbler**.

P.15. Kidasi 34°49'11.1"N 32°42'25.7"E (34.819755, 32.707151) From the old Limassol road just east of Asprokremmos Dam, take the road for 'Troodos, Agios Nikolaos' etc., drive 18km along the Diarizos Valley and pass between two huge boulders ('Rock of Chasampoulion') and a short stretch of riverine forest. The mature alders here host **Eurasian Siskin** in winter and **Cyprus Scops Owl** calls in spring at dusk and dawn.

Drive a further 2km to Kidasi, cross the river, and take the graded track along the north flank of the hill, to look for **Finsch's Wheatear** in winter. Scan the sky for **Griffon Vulture, Bonelli's Eagle, Long-legged Buzzard** and **Peregrine Falcon**, which patrol throughout the year.

P.16. Mavrokolymbos Dam 34°51'27.0"N 32°24'22.0"E (34.857500, 32.406111) A small dam with steep, gorse-covered banks and small numbers of waterfowl. Signposted 1km south of the turn-off for Pegeia on the Paphos–Coral Bay road, drive 2.5km on a surfaced road to the dam. In spring and summer this is a fairly reliable site for **Cyprus Wheatear, Cyprus Warbler** and **Cyprus Scops Owl**.

P.17. Nata (Choletria) Bridge 34°46'30.0"N 32°34'49.0"E (34.775000, 32.580278) The Ezousas River crossing east of Nata village, with pools and reeds; a pleasant place to birdwatch from the bridge Feb–May. Likely species include **Bonelli's Eagle, Little Ringed Plover, Eurasian Stone-curlew** and **Red-rumped Swallow**.

Akrotiri Peninsula and Limassol District

A.1. Zakaki Marsh 34°38'32.8"N 33°00'12.4"E (34.642444, 33.003444) From the A6 motorway west of Limassol, take the turn-off to 'Ypsonas, Port'. After leaving the motorway, travel 6km following signs for 'Lady's Mile'. The reedbed and Zakaki Marsh tower hide are on the right 200m after the container port roundabout. Best Aug–May, the condition of the reeds varies and they are sometimes overgrown, making views difficult from the hide, but there are open pools on the roadside and to the left beyond the hide, enabling good views of herons, crakes, waders, warblers and other species. It is a regular roost for waterbirds, and a bathing site for gulls. **Eurasian Reed** and **Cetti's Warblers** breed in good numbers, and **Ferruginous Duck** and **Water Rail** are present most of the year. In autumn birds of prey fly over, including **Eurasian Honey Buzzard, Western Osprey, Black Kite, Marsh, Montagu's** and **Pallid Harriers**. In spring and autumn a variety of wetland species occurs, including **Black-crowned Night** and **Purple Herons, Great Egret, Glossy Ibis, Spotted, Little** and **Baillon's Crakes, Collared Pratincole, Citrine Wagtail, Bluethroat, Moustached Warbler** and **Penduline Tit**. Some of these overwinter and several also occur in spring. **Little Bittern** is regular in late spring and occasionally breeds. Rarities seen here in recent years include **White-headed Duck, Striated Heron, Black-winged Pratincole, Little Swift, White-throated Kingfisher, Pied Kingfisher, Bearded Reedling** and **Common Rosefinch**.

The Zakaki Marsh hide overlooks the main reedbeds, where Penduline Tit, Moustached Warbler and Bluethroat are regular in winter (Colin Richardson).

A.2. Lady's Mile 34°38'26.1"N 33°00'27.4"E (34.640583, 33.007611) The start of this stretch of seasonal saltmarsh lies 300m past Zakaki Marsh. Driveable tracks run the length of the site, parallel to Limassol Bay, stretching for c.4km. Best Sep–May. The shallow saline pools, bordered by salt bush, support only a limited number of species most of the year, mainly **Kentish Plover**, **Little Stint** and **Dunlin**, but after spring rains and during high tides there can be influxes of waders including **Black-winged Stilt**, **Ruff**, **Temminck's Stint** and **Curlew Sandpiper** from the nearby salt lake. Large numbers of loafing gulls occur from late Oct, the most interesting being **Slender-billed**, **Armenian**, **Caspian** and **Pallas's Gulls**. Many rest on the sea, scavenging in front of the Oasis Restaurant. **Sandwich Tern** is the most likely tern in winter, with **Gull-billed** and **Little Terns** in spring. On good migration days you might see **Greater Sand Plover**, **Broad-billed Sandpiper** or **Red-necked Phalarope**. Rarities seen here include **Black-throated Diver** (on Limassol Bay), **Bar-tailed Godwit**, **Red Knot** and the country's only **Baird's Sandpiper** (winter 2009).

A.3. Akrotiri Salt Lake 34°36'02.3"N 32°57'37.4"E (34.600639, 32.960389) The salt lake can be viewed using a telescope from the balcony of the Akrotiri Environmental and Educational Centre, on the south-west side of the lake. It is difficult to approach the lake closely as most tracks are too soft or dangerous to drive or walk after rain, but there are some well-used tracks on the east side (accessed from Lady's Mile). Best late Jul–May, dependent on water levels. The surrounding salt flats are interspersed by low dunes covered with salt bush and tamarisk. There is year-round inflow of brackish water from Zakaki Marsh channel. The lake is well known for its wintering **Greater Flamingo** population, but also hosts nesting **Kentish Plover** and **Black-winged Stilt**. Waterfowl of all kinds occur, notably **Ruddy Shelduck**, **Common Crane**, **Collared Pratincole**, **Broad-billed Sandpiper**, **Marsh Sandpiper**, **Red-necked Phalarope** and **Caspian Tern**. It is the only regular site in Europe where groups of **Demoiselle Crane** stopover on autumn migration. They can be seen far out on the dry salt flats on several mornings late Aug–mid Sep, before using thermals to depart 08.30–10.00h. **Eleonora's Falcons** are seen hunting Jun–Oct, and migrating raptors, including **Eurasian Honey Buzzard**, **Black Kite** and harriers, are frequent Sep–Oct. Recent rarities include **Crested Honey Buzzard**, **Red-necked Grebe** and **Black-winged Pratincole**.

Greater Flamingos are regular at the Akrotiri Salt Lake for most of the year (Colin Richardson).

The tower hide at Akrotiri Marsh gives great views over the Phasouri reedbeds (Colin Richardson).

A.4. Akrotiri Marsh (Phasouri Reedbeds) 34°37'58.2"N 32°56'03.0"E (34.632833, 32.934167) A Ramsar site, Important Bird Area (IBA) and Special Protection Area (SPA), this is a very large area of reeds; restoration of the site was completed in 2017. From Kolossi village drive south towards Akrotiri, then, after 4.5km, turn right at the Malcolm Cat Sanctuary; after 600m turn sharp right to reach the site. Access to the North (tower) Hide is on the left; 800m further along the paved road you will come to the Ground Hide on the left, reached by a screened walkway. This can be quite productive for herons, crakes, waders and other reed-dwelling species, particularly in spring. However in recent years, possibly due to water supply, numbers have fallen. In winter **Marsh Harrier** is common and you might see or hear **Bluethroat**, **Penduline Tit** and **Moustached Warbler**. The reeds are home to several pairs of **Ferruginous Duck**, while **Little Bittern** occurs in late spring and may breed occasionally. Hirundines and swifts are regular and in autumn **European Bee-eater** is abundant. Sep–Oct are best to see migrating raptors, especially harriers, eagles and falcons. The island's **Eleonora's Falcon** colonies are nearby, and birds often visit the reedbeds to hunt. In winter **Common Kingfisher** is sometimes joined by **White-breasted** and **Pied Kingfishers**. Other good birds in recent years include **Ruddy Shelduck**, **Black Stork**, **Black-winged Kite**, **Black-winged Pratincole**, **Terek Sandpiper**, **Red Phalarope**, **Common Crane**, **Levant Sparrowhawk**, **Saker**, **Little Swift**, **Blue-cheeked Bee-eater**, **Richard's Pipit**, **Caspian Stonechat**, **Paddyfield Warbler**, **Red-breasted Flycatcher**, **Bearded Reedling**, **Rose-coloured Starling**, **Dead Sea Sparrow** and **Trumpeter Finch**.

A.5. Bishop's Pool 34°35'50.7"N 32°59'08.4"E (34.597417, 32.985667) The pool lies in the centre of an area of private farms bordering Akrotiri airfield. The fenced area is accessed via a gate, adjacent to the Holy Monastery of St Nicholas of the Cats. It is usually unlocked and you should close it after you enter/leave. The pond attracts a variety of waterfowl and herons, including resident **Ferruginous Duck**, throughout the year and raptors regularly patrol the area. Passage migrants are attracted to the fields, hedges and tall trees. Rare visitors in past years have included **White-headed Duck**, **Levant Sparrowhawk**, **Pied** and **White-throated Kingfishers**, **Yellow-browed Warbler** and **Red-breasted Flycatcher**.

A.6. Akrotiri Gravel Pits 34°36'19.9"N 32°56'10.6"E (34.605528, 32.936278) An expanse of gravel plains and saturated salt flats, dotted with security masts. It can seem bird-less at times, but over the years has produced some excellent records. Best in spring. The focal point for small migrants in spring is the Agios Georgios chapel garden. West of the chapel is a large brackish pool, which should be checked for waders. North of the chapel, the open area of 'pits' or gravel plains, dotted with brackish pools and an area of sand dunes to the south-west, are worth checking in spring for larks, wheatears, pipits and warblers, while in autumn raptors drift over. In Curium Bay hundreds of migrant ducks congregate in spring before coasting west or flying over the island at night. Recent rarities have included **Red-breasted Merganser**, **Demoiselle Crane**, **Sociable Plover**, **Cream-coloured Courser**, **Black-bellied Sandgrouse**, **Bar-tailed** and **Bimaculated Larks**, **Hooded Wheatear**, and **Turkestan** and **Southern Grey Shrikes**.

A.7. Kensington Cliffs 34°40'00.4"N 32°51'44.3"E (34.666780, 32.862301) From Kolossi drive west (from the Akrotiri Peninsula take the M1 road) then climb the cliff road past Kourion, to the Episkopi Garrison, in the UK Sovereign Base; 200m past the Sanctuary of Apollo Hylates, turn left to a viewpoint on the sea cliffs overlooking Quarry Beach (Zapallo Bay). This is a key site for Cyprus' remaining **Griffon Vultures**. Best in spring and summer when **Eleonora's Falcon** and **Alpine Swift** are also present. The cliffs also host **Peregrine Falcon**, **Cyprus Wheatear** and **Cyprus Warbler**, the latter now in serious decline. In winter **Wallcreeper** and **Blue Rock Thrush** may also be found with luck.

Kensington Cliffs provide dramatic views of resident Griffon Vulture, and of visiting breeders Eleonora's Falcon and Alpine Swift (Colin Richardson).

A.8. Pissouri Bay, East Cliffs 34°39'32.1"N 32°44'03.3"E (34.658917, 32.734250) Best Sep–Nov and Mar–May. The area of mixed scrub and woodland on the cliffs and inland to the east of Pissouri Bay is rich in species, including all endemics. Visiting breeders include **Eleonora's Falcon**, **Alpine Swift**, **European Roller**, **European Nightjar** and **Great Spotted Cuckoo**. In winter **Finsch's Wheatear** occurs and **Wallcreeper** can be found on the west cliffs. Recent rarities include **White-tailed Plover**, **Little Swift** and **Cinereous Bunting**.

A.9. Germasogeia Dam (and Finikaria) 34°44'39.0"N 33°04'59.5"E (34.744168, 33.083196) The area of the dam includes the river inflow and, when levels are high, it attracts many birds Sep–May. The dam wall often hosts **Blue Rock Thrush** in winter. Aside from interesting migrant herons, waterfowl on the dam is usually limited to common species. The shallow end can be reached by driving 4km to Finikaria, through the village to the water's edge and reedbed via graded tracks (34°45'29.3"N 33°05'32.6"E [34.758140, 33.092380]). In winter and spring, when the water level is high, the reedbeds are attractive to **Little** and **Spotted Crakes**, **Jack Snipe**, **Bluethroat** and **Moustached Warbler**, while **Bonelli's Eagle** and **Long-legged Buzzard** are regular overhead. Small groups of **Eurasian Crag Martin** often visit Nov–Mar and **European Nightjar** hunts over the reeds in late spring. **Pied Kingfisher** has turned up several times.

Larnaca District

L.1. Larnaca Salt Lakes 34°52'52.5"N 33°36'59.9"E (34.881250, 33.616639) The salt lake is seasonal and relies on rainfall Oct–May; best viewed from the Tekke Mosque road or from the Aphrodite Trail car park, 2km north of the old airport roundabout. The salt lake is well known for its confiding **Greater Flamingo** population, with many **Common Shelduck** in winter and **Black-necked Grebe** in spring. There is often a selection of gulls, including **Slender-billed, Caspian, Heuglin's** and occasionally **Mediterranean** and **Pallas's**. Other recent visitors include **Common Crane, Greylag Goose, Eurasian Spoonbill** and **Pied Avocet**. Other waders, mostly **Kentish** and **Ringed Plovers, Little Stint** and **Dunlin**, feed around the lake. The Larnaca Airport Pools North lie adjacent to the airport on the south side of the old airport road, and host sea ducks, waders, gulls and, in recent years, **Red-necked Phalarope** in spring. The north end of the lake is accessed from Kamares Aqueduct 34°54'43.5"N 33°35'55.5"E (34.912070, 33.598743) beside the B5 main road to the Paphos/Nicosia motorway junction. The Salt Lake Hiking trail runs along the east side of the lake, past a brackish marsh which often has **Jack Snipe, Temminck's Stint** and **Marsh Sandpiper** in winter and spring, and the reeds have **Penduline Tit** in winter.

L.2. Larnaca Sewage Works area 34°51'37.3"N 33°37'37.6"E (34.860361, 33.627111) The area of fields and wetlands adjacent to the sea south of Larnaca airport is one of the island's top birdwatching sites. From the airport direction, drive to the village of Meneou, then bear left at the F401 ('To the Beach') and head to the sea. As you approach the T-junction after 2.8km, you will find Meneou Pools on the right and the vast Airport Pools South on the left – both are worth checking. Turn left here, and the Larnaca Sewage Works hide is visible on the left after 2km, accessed by a short track. The two lined pools of recycled water visible from the tower hide are interesting year-round, but best Aug–May. Among the standard waterfowl, waders and gulls, **Red-breasted Goose, Marbled** and **White-headed Ducks, Terek Sandpiper, Audouin's Gull** and **Caspian Tern** have occurred. **Black-headed Wagtail** and **Spectacled Warbler** breed in the salt bush; the fields and scrub have produced **Caspian** and **Pacific Golden Plovers, Lesser Short-toed** and **Bar-tailed Larks, Hooded Wheatear** and **Asian Desert Warbler**. **Eurasian Stone-curlew** nests nearby and roosts in the open scrub. Elsewhere, Spiros Pools (between the sea and sewage works), Meneou Pools and

The tower hide at the Larnaca Sewage Works overlooks two large settlement pools, which host hundreds of wintering waterfowl (Colin Richardson).

the Airport Pools South have held **Ruddy Shelduck**, **Collared Pratincole**, **Red-necked Phalarope** and **Broad-billed Sandpiper**. Rarities have included **Black-winged Pratincole** and **Grey Phalarope**. **Little** and **Common Terns** arrive in Apr and occasionally breed. Check the beach and shoreline for waders and gulls including **Greater Sand Plover**, **Sanderling**, **Eurasian Curlew** and **Ruddy Turnstone**. A **Franklin's Gull** was at the sewage pools Jul 2006 and a **White-eyed Gull** on the beach in Jun 2017. **European Shag** and **Sandwich Tern** often rest on offshore buoys. The trees around the holiday homes and caravans have yielded **Red-breasted Flycatcher**, **Thrush Nightingale** and **Semi-collared Flycatcher**. On the coastal track, past the desalination pump station, the large field on the left formerly had breeding **Calandra Lark**, but depending on the crop can still host interesting birds and often attracts raptors.

L.3. Oroklini Lake 34°58'14.4"N 33°39'17.2"E (34.970679, 33.654771) A Special Protected Area (SPA) with two tower hides and information boards. Located 5km north of Larnaca, Oroklini (or 'Voroklini') is signposted, and the main hide is 1km inland of the coast road, visible from the Larnaca–Agia Napa motorway. **Ferruginous Ducks** nest most years and **Red-crested Pochard** for the first time in 2009. **Cattle Egrets** colonised in 2014, and hundreds roost in the late afternoon. Small groups of **Greater Flamingo** are a major attraction for visitors, while the brackish channels to the south host breeding **Black-winged Stilt** and occasionally **Little Tern**. **Spur-winged Lapwings** are noisy residents. Large numbers of waterfowl overwinter, sometimes including **Black-necked Grebe**. Assorted gulls, including **Pallas's**, **Slender-billed** and **Mediterranean Gull**, often visit to bathe. The marsh attracts the first swifts and hirundines of spring. Regular herons include **Great Egret** and **Purple**, **Squacco** and **Black-crowned Night Herons**, plus **Little** and **Eurasian Bitterns**; and visiting waders include **Pied Avocet**, **Collared Pratincole**, **Jack Snipe**, **Temminck's Stint**, **Spotted Redshank**, **Black-tailed Godwit** and **Marsh Sandpiper**. Raptors are regular too, the commonest being **Marsh Harrier**. The reedbeds host wintering **Bluethroat**, **Moustached Warbler**, **Penduline Tit** and **Water Rail**, and **Citrine Wagtail**, **Great Reed Warbler**, and **Little** and **Spotted Crakes** occur on passage (some of these near the south hide). Less common on passage are **Ruddy Shelduck**, **Great White Pelican**, **Black** and **White Storks**, **Caspian Tern**, **Blue-cheeked Bee-eater** and **Rose-coloured Starling**. Rarities in recent years have included **White-headed Duck**, **Terek Sandpiper**, **Pin-tailed Sandgrouse**, **Pied Kingfisher**, **Little Swift**, and **Asian Desert** and **Yellow-browed Warblers**.

The tower hide at Oroklini Lake overlooks the main marsh (Colin Richardson).

Cape Greco is one of the country's top migrant hotspots in spring (Colin Richardson).

L.4. Cape Greco 34°58'13.9"N 34°04'12.5"E (34.970528, 34.070139) From Agia Napa head east for 7km, following signs for 'Kavo Gkreko'. As you reach Cape Greco, before turning right at Louis's hamburger van, search the young pine plantations on both sides of the road for migrants. Continue downhill past the Environmental Centre and after 300m turn left towards Agioi Anargiroi church, and search the area including the sea, where **Audouin's Gull** and other passing seabirds are regular. Return to the junction and drive towards the high masts at the tip of the cape. This large area, including the south cliffs below the army base, the bushes on the north side (Blue Lagoon) and the Sea Caves area, all shelter migrants. The whole site is good year-round, especially Mar–May, but is quieter in autumn, apart from migrant raptors and coastal migrants. Key species include **Cyprus Warbler**, **Cyprus Wheatear**, **Lesser Short-toed Lark**, **Blue** and **Common Rock Thrushes**, **Finsch's Wheatear**, **Subalpine** and **Icterine Warblers**, **Semi-collared Flycatcher**, **Rose-coloured Starling**, and **Cretzschmar's**, **Ortolan** and **Black-headed Buntings**. The long list of recent rarities includes **Hypocolius**, **White-throated Robin**, **Caspian Stonechat**, **Kurdish**, **White-crowned** and **Desert Wheatears**, **Olive-tree**, **River**, **Ménétries's** and **Asian Desert Warblers**, **Red-breasted Flycatcher**, **Brown-necked Raven**, **Steppe Grey Shrike**, **Isabelline** and **Red-tailed Shrikes**, **Desert** and **Trumpeter Finches**, **Pale Rockfinch**, **Rock Sparrow** and **Cinereous Bunting**.

L.5. Agia Napa Sewage Works (Paralimni–Agia Napa Water Treatment Plant, WTP) 34°59'26.3"N 34°02'23.8"E (34.990712, 34.039939) From the Agia Napa roundabout drive towards Cape Greco and after 3.2km turn left at the road for Agia Paraskevi chapel. This surfaced road heads towards the gates of the WTP, but after 1.1km turn left onto a very rough track. After 350m, turn right and you should arrive at a point where you can see the ponds through the fence. At the tall rocks on your left, **Common** and **Blue Rock Thrushes** are regular. In spring and autumn the area of scrub above and below the WTP is rich in shrikes, wheatears, warblers and buntings. In late summer (Jul–Sep) migrants have included **Barred** and **Eastern Orphean Warblers**, **Rufous-tailed Scrub Robin** and **Rose-coloured Starling**.

L.6. Limnara Beach 34°58'23.1"N 34°02'12.4"E (34.973088, 34.036786) An interesting stretch of rocky coastline and scrub formerly known as Kermia Beach. Drive 4km from the Agia Napa roundabout towards Cape Greco, and turn sharp right at a road that takes you back on a parallel road to Agia Napa. Look for the first easy track on the left towards the beach (a new housing development after 150m may give access to the beach, or a farm track after 500m) and follow the coastal track east until you overlook a couple of rocky promontories at Limnara Beach, which is a favoured roosting site for **Greater Sand Plover** and **Audouin's Gull**, while **Gannet** and **Red-breasted Merganser** have been seen offshore. Back on the main Agia Napa–Cape Greco road, 50m before the sharp right turn-off, on the left (north side) of the road there is a pull-in and a short track. Walk the track for 50m and through the bushes to a hidden pool with reeds, known as Agia Paraskevi pond, which is good for herons and crakes, and the bushes attract migrants.

L.7. Achna Dam 35°03'20.0"N 33°48'49.9"E (35.055556, 33.813861) A medium-sized storage reservoir with bays, inlets and a shallow end with grasses and reeds. A good selection of species Sep–May, especially in spring. Head east from Larnaca on the Agia Napa motorway and take the turn signed 'Xylotymvou, Achna' etc.; after another 10km, turn right at a big sign 'Akhna Dam'. The dam can be viewed from either side of the dam wall, and is best explored on foot using the fishermen's tracks around the perimeter. The quiet shallow end to the west produces the most interesting species. In winter you often find only **Cattle Egrets**, Coots, cormorants and grebes due to the presence of fishermen, but interest picks up Mar–May, with **Little Bittern**, **Eurasian Spoonbill**, **Ruddy Shelduck**, **Red-crested Pochard**, and **White-winged** and **Whiskered Terns** all possible. Specialties in recent years include **Great White Pelican**, **Black Stork**, **Jack Snipe**, **Moustached Warbler** and **Penduline Tit**, and surprises are frequent. Groups of **Common Crane** fly over in spring and migrant raptors, including **European Honey Buzzard**, **Booted Eagle**, **Pallid** and **Montagu's Harriers**, **Western Osprey** and **Red-footed Falcon**, occur in autumn. Rarities, including **Black-winged Pratincole**, **Pectoral Sandpiper**, **Caspian Tern**, **Pied Kingfisher** and **Buff-bellied Pipit**, have occurred multiple times.

Achna Dam is a small but very productive reservoir attracting herons, pelicans, waders, raptors and kingfishers throughout the year (Colin Richardson).

Secondary sites in Larnaca District

L.8. Pervolia–Kivisili–Mazotos–Petounta Vast cereal fields and rocky coastline, plus seasonal marsh and headland at Petounta near the housing development and church (signposted Panageia Petounta 34°47'10.5"N 33°29'41.1"E [34.786252, 33.494758]). Best Mar–May. Species include **Pallid** and **Montagu's Harriers**, **Red-footed Falcon**, **Lesser Kestrel**, **Eurasian Stone-curlew**, **Cream-coloured Courser**, **Bimaculated Lark** and **Calandra Lark** (breeds). Recent rarities have included **Namaqua Dove**, and **Turkestan**, **Steppe Grey** and **Southern Grey Shrikes**.

L.9. Kiti Dam 34°52'08.6"N 33°33'33.4"E (34.869063, 33.559271) Accessed by a farm track from Tersefenou (2km) or by driving north through the suburbs of Kiti (2.5km) The dam is adjacent to Agios Georgios Arperas chapel. Best late Mar–early May. When not dry the dam hosts a good selection of waterbirds, including ducks, herons, crakes and waders. In drought years the woodland and reeds at the base of the dam wall can be quite productive in spring. Look for roosting **Long-eared Owl**, plus **Great Spotted Cuckoo**, **Eastern Bonelli's** and **Wood Warblers**, **Pied**, **Collared** and **Semi-collared Flycatchers**, **Golden Oriole** and shrikes.

L.10. Paralimni Lake 35°01'37.9"N 33°58'20.0"E (35.027194, 33.972222) This is the country's third salt lake, and can be productive Aug–May, but is very dependent on rainfall and is often dry (although Sotira Pools on the north-west side retain water most of the year). It is accessed from the south end of the Paralimni–Sotira road (near the Shooting Club), from where you can walk out on the salt flats. This area can hold hundreds of waterbirds, often visible through the heat haze, and there are likely to be some surprises. Check the brackish channels, muddy pools and reeds, which always host migrants. On a good day species might include **Little** and **Eurasian Bitterns** (winter), **Great Egret**, **Pallid Harrier**, **Common Crane**, **Spotted** and **Little Crakes**, **Greater Sand Plover**, **Pied Avocet**, **Temminck's Stint**, **Great** and **Jack Snipes**, **Red-necked Phalarope**, **Pallas's Gull** (winter), **Whiskered** and **White-winged Terns** (autumn), **Bimaculated Lark**, **Red-throated Pipit**, **Citrine Wagtail**, **Bluethroat** (winter) and **Eastern Bonelli's Warbler**. Recent rarities have included **Ruddy Shelduck** (winter), **Saker**, **Baillon's Crake**, **White-tailed Lapwing**, **Eurasian Dotterel** and **Caspian Plover**.

Troodos Forest 34°55'18.8"N 32°52'40.5"E (34.921889, 32.877917)

A birdwatching trip to Cyprus is not complete without a visit to Troodos and its surrounding forest. The cool air and complete change of habitat is refreshing, and the prospect of finding some new species exciting. Start at Troodos village (altitude 1,750m) where the endemic subspecies **Cyprus Coal Tit**, **Cyprus Jay** and **Dorothy's Short-toed Treecreeper**, and regional endemic subspecies **Cyprus Red Crossbill** and **Cyprus Wren** are present year-round. Red Crossbill is generally only present above 1,100m where it feeds on the seeds of Black Pine *Pinus nigra*. In spring and summer **Masked Shrike** is fairly common, while a sizeable **Pallid Swift** colony is present in Troodos village. **Northern Goshawk** is an elusive breeder in remote areas, **Grey Wagtail** breeds at waterfalls and along fast-flowing perennial streams, while **Cyprus Wheatear** is common everywhere. **Eurasian Blackbird** has only recently colonised the forest but is already locally abundant. Other forest breeders include **Cyprus Scops Owl**, **Common Cuckoo**, **Eurasian Hoopoe**, **Woodlark**, **Eurasian Crag Martin**, **Common Nightingale**, **Spotted Flycatcher**, **European Serin**, **Common Chaffinch** and **Cretzschmar's Bunting**. Many winter visitors are subject to weather-related influxes, with **Ring Ouzel** and **Redwing** the least common, while **Dunnock**, **Fieldfare**, **Mistle Thrush**, **Goldcrest**, **Eurasian Siskin**, **Hawfinch** and **Brambling** may not be star birds for European visitors, but do impress the locals! The rare **Pine Bunting** is sometimes found in the highest forest areas, often with **Yellowhammer**, another uncommon winter visitor.

Troodos Forest, near the Persephone Trail, is dominated by Black Pine woodland (Colin Richardson).

T.1. Livadi Tou Passia 34°56'04.8"N 32°53'56.5"E (34.934667, 32.899028) 2.5km downhill from Troodos on the Nicosia road, a picnic site set around a wet meadow. In spring and summer you should find all of the endemic forest subspecies here, in addition to visiting Masked Shrike, Spotted Flycatcher and Eurasian Hoopoe. In winter, thrushes, finches and other specialities occur here too, plus Pine Bunting. There are seldom many picnickers so it is always quiet.

T.2. Almirolivado Picnic Site (Giant Junipers) 34°55'49.6"N 32°54'09.0"E (34.930444, 32.902500) 3.5km downhill from Troodos on the Nicosia road. Open areas of pine, juniper and cedar, with grassy marsh in winter. All endemics occur, plus Redwing, Fieldfare, Brambling, Hawfinch, Yellowhammer and Pine Bunting in winter.

T.3. Troodos Botanical Garden, Pano Amiantos 34°55'48.6"N 32°54'50.9"E (34.930167, 32.914139) 6km downhill from Troodos on the Nicosia road. Formal botanic garden, with mix of coniferous and broadleaf woodland, and panoramic views. Notable species include Long-legged Buzzard, Blue Rock Thrush, Hawfinch, European Serin and Eurasian Siskin. A reliable site in summer for Eurasian Crag Martin.

T.4. Platania Forest station 34°56'57.5"N 32°55'40.2"E (34.949306, 32.927833) 10km downhill from Troodos on the Nicosia road. Offices of the Forestry Department with access roads and walking tracks. Broadleaf and coniferous forest plus a tree nursery, which hosts Dunnock, Goldcrest, Brambling and Hawfinch in winter, as well as most of the endemics (except Red Crossbill). Yellow-browed Warbler is regular in late autumn.

T.5. Kelephos Bridge and Arminou Dam (shallow end) 34°53'19.7"N 32°44'54.3"E (34.888806, 32.748417) Best spring and summer. Signposted from the village of Agios Nikolaos, and reached after 6.5km by a good asphalt road that gives views of Arminou Dam before it plunges into Paphos Forest. The Diarizos River runs throughout the year under this medieval Venetian bridge. Walk upstream to search for

Kelephos Bridge spans the Diarizos River, where Common Nightingale, Spotted Flycatcher, Grey Wagtail and several Cyprus forest endemics breed nearby (Colin Richardson).

Grey Wagtail, Cyprus Wren, Cyprus Wheatear, Spotted Flycatcher, Eastern Olivaceous Warbler, Common Nightingale, Cyprus Jay, Short-toed Treecreeper and Cyprus Coal Tit. Northern Goshawk is regular too. The shallow end of Arminou Dam is c.800m downstream of the bridge, accessed by a rough track from the signposts near Kelephos Bridge. As you walk towards the dam you are likely to see many birds possibly including Black-headed and Cretzschmar's Buntings, plus migrant sandpipers and herons.

T.6. Agios Nikolaos to Omodos vineyards 34°51′51.6″N 32°47′20.7″E (34.864321, 32.789079) Best late Apr–Jul. A 5km mountain road with spectacular views, linking the two villages, while passing through extensive vineyards. The area has large populations of Black-headed Bunting and Cyprus Wheatear, while also likely are Hobby, European Bee-eater, Great Spotted Cuckoo, European Turtle Dove, Red-rumped Swallow, Eastern Olivaceous and Cyprus Warblers, Masked Shrike, Spanish Sparrow, European Serin and Cretzschmar's Bunting.

Nicosia District

N.1. Athalassa Park 35°07′31.7″N 33°23′05.1″E (35.125472, 33.384750) One of the best birdwatching sites in south Nicosia. Leave the A1 motorway on the B22 east, then go straight (second exit) at the hospital roundabout (towards Geri) and after 600m turn left into Athalassa Park; park in car park no.2. The lake is c.100m east of the car park. There is a woodland walk around the lake, with a good selection of waterbirds most of the year and a water treatment pool (viewable on the north side) holds grebes and ducks. The main lake hosts a Cattle Egret colony in spring and summer. There are various corners, reedy islands and open water visible from the main path, and a raised hide on the south-east side where you can watch without causing disturbance. Species include Black-necked Grebe, Little Bittern, Great Egret, Black-crowned Night Heron, Eurasian Spoonbill, Ferruginous Duck, Jack Snipe, White-winged and Whiskered Terns, Pied Kingfisher, Great Spotted Cuckoo, European Roller, European Bee-eater, and Eastern Olivaceous and Cyprus Warblers. Long-eared Owl breeds in good numbers in the wooded areas.

N.2. Agios Sozomenos 35°03′09.3″N 33°26′20.9″E (35.052583, 33.439139) This extensive area of farmland, cliffs and ruins forms part of an abandoned Turkish Cypriot village community, and is 15km south-east of Nicosia just east of Dali near the 'buffer zone'. Long-legged Buzzard, Bonelli's Eagle and Peregrine Falcon breed locally, as do Cyprus and Spectacled Warblers. It's a good area for winter specialties, including **Finsch's Wheatear** and **Blue Rock Thrush**. **Hen** and **Marsh Harriers** regularly patrol the fields, where **Cattle Egret**, **Red-throated Pipit**, **European Serin**, **Corn Bunting** and occasionally **Calandra Lark** can be found. Breeders include **Little**, **Barn** and (formerly) **Long-eared Owls**, **Eurasian Stone-curlew**, **European Roller** and **Cyprus Wheatear**. A water treatment pool north of the site attracts herons and waders. Rarities have included **Olive-backed Pipit**.

Northern Cyprus

Nth.1. Kioneli Dam (Gönyeli Reservoir) 35°14′01.3″N 33°18′17.6″E (35.233694, 33.304889) Best Nov–Mar. About 5km north-west of Nicosia, accessed from the Nicosia–Morphou (Lefkoşa–Güzelyurt) highway. Take the turn for Kioneli (Gönyeli), just west of the Kyrenia (Girne) roundabout. Drive through the village and after 2.6km take a right turn signposted 'Le Chateau Romance'. Follow the winding road for c.1.6km, across open land, go under a concrete road bridge, and you will emerge at Le Chateau Romance restaurant. Follow the bumpy track over the low hill, and you will see the reservoir, which in winter hosts hundreds of waterfowl. From late autumn to spring there are also variable numbers of **Black-necked Grebe**, **Northern Shoveler**, **Ruddy Shelduck**, **Ferruginous Duck** and **Great Egret**, plus **Water Rail**, **Spur-winged Lapwing**, **Black-winged Stilt**, **Spotted Redshank**, **Marsh Sandpiper**, and **Whiskered** and **White-winged Terns**. There are interesting passerines too, such as **Moustached Warbler**, **Penduline Tit**, **Bluethroat**, **Woodlark** and **Water Pipit**.

Kioneli Dam is located in the northern suburbs of Nicosia, and hosts an impressive selection of waterfowl (Colin Richardson).

Nth.2. Kanli Dam (Kanlıköy Reservoir) 35°14′16.6″N 33°15′34.9″E (35.237944, 33.259694) Best Nov–Mar. A medium-sized dam which can attract large numbers of winter waterfowl. Although just 4km west of Kioneli (Gönyeli) Dam, it is best reached from the main Nicosia–Morphou (Lefkoşa–Güzelyurt) highway. Take a turn after Kioneli, follow a service road that passes through the small village of Kanli (Kanlıköy) and continue north on a narrow road and track for 2.5km. The list of birds recorded includes **Great Crested Grebe**, **Garganey**, **Gadwall**, **Common** and **Red-crested Pochards**, **Squacco Heron**, **Great Egret**, **Little Crake**, **Spur-winged Lapwing**, **Marsh Sandpiper** and **Eurasian Stone-curlew**. Noteworthy sightings include **Demoiselle Crane** (spring) and **Goosander** (winter).

The Mia Milia Water Treatment Plant hosts thousands of waterbirds in winter, and is dominated by the dramatic Kyrenia Mountains (Colin Richardson).

Nth.3. Mia Milia Water Treatment Plant (Haspolat WTP) 35°11'35.1"N 33°25'43.5"E (35.193083, 33.428750) Good all year. Located c.7km east of Nicosia (Lefkoşa) along the Nicosia–Famagusta (Lefkoşa–Gazimağusa) highway. Take the turn-off at the roundabout (note the large mosque) for Mia Milia (Haspolat) and head south for 1.5km to a T-junction. Turn right and after 1.8km turn left, then follow the winding road for 2.2km to the entrance. As a courtesy, make yourself known to the gateman and explain you are a birdwatcher. This sewage plant is of the 'traditional' type consisting of vast settlement pools and soakaways spread over 2–3 km². The settlement pools are divided by reed-lined embankments and host thousands of waterbirds in Oct–Mar. Groups of **Ferruginous Ducks** are regular, large numbers of **Spur-winged Lapwings** resident and **Cattle Egret** occurs year-round. Other interesting species to look for include **Black-necked Grebe, White-headed Duck, Greater Flamingo, Great White Pelican, Eurasian Spoonbill, Common Shelduck, Garganey, Red-crested Pochard, Steppe Buzzard, Collared Pratincole, Spotted Redshank, Marsh Sandpiper, Jack** and **Great Snipes, Black-tailed Godwit**, and **Whiskered** and **White-winged Terns**. Five soakaway/slurry beds west of the site are surrounded by reeds and require close inspection because they usually host many migrants, notably crakes, **Citrine Wagtail, Bluethroat, Moustached** and **Great Reed Warblers**, and **Penduline Tit**.

Nth.4. Trachoni/Palaikythro Pools and fields (Demirhan/Balıkesir Pools) 35°11'32.6"N 33°28'28.3"E (35.192389, 33.474611) Best Sep–May. Drive 10km east from Lefkosia on the Lefkosia–Famagusta (Lefkoşa–Gazimağusa) highway to the Neo Khorio (Minareliköy) roundabout (also the junction from Kyrenia/Girne to the north) and take the exit south, which soon becomes a rough gravel track, though quite suitable for most cars. Continue south between irrigated fields for 2.5km to a wooden bridge over the Pedieos River (Kanli Dere), the longest river in Cyprus and partly fed by the Mia Milia WTP upstream. Over the bridge, turn left and the reedy Trachoni (Balıkesir) Pools appear after 250m. These pools are seasonal, but when in good condition in Nov–May there are waterfowl, waders and others, including **Ferruginous Duck, Garganey, Little Bittern, Little** and **Spotted Crakes, Spur-winged Lapwing, Black-winged Stilt, Great Snipe,**

Bluethroat, **Moustached Warbler**, **Penduline Tit** and **Common Reed Bunting**. Tour the irrigated fields in season (spring is best) and you should find **Cattle Egret**, **Red-footed Falcon**, **Lesser Kestrel**, **Black Francolin**, **European Bee-eater**, **Alpine Swift**, **Calandra Lark** (breeds locally), **Red-throated Pipit**, **Northern Raven** and harriers. Many other herons, raptors, waders, larks, pipits, wheatears and **Corn Bunting** regularly occur. Rarer species such as **White Stork** and **Short-toed Snake Eagle** have also occurred.

Nth.5. Agios Loukas Freshwater Lake South (Ayluga South) 35°07′22.1″N 33°54′39.4″E (35.122806, 33.910944) Located 1km west of Famagusta, on the south side of the main Famagusta–Nicosia (Gazimağusa–Lefkoşa) highway. On entering the highway from the roundabout, drive 900m and turn left at a track through eucalyptus woodland. The lake and pools are visible immediately. The lake can also be accessed from the nearby village of Agios Loukas, which backs onto the lake. Water levels vary; high levels attract wintering ducks, and low levels expose islands and shorelines for waders. The lake hosts a large colony of **Cattle Egret** in Mar–Jul, and the activity has attracted a growing colony of **Glossy Ibis**, plus nesting **Squacco** and **Black-crowned Night Herons**. Other waterbirds are also regular, including **Little Bittern**, **Great White Pelican**, **Ferruginous Duck**, **Little** and **Spotted Crakes**, **Pied Avocet**, **Eurasian Spoonbill**, **Spur-winged Lapwing** (breeds), **Great Snipe**, **Whiskered Tern** and **Bluethroat**. Raptors are regular too, notably **Black Kite**, **Western Osprey**, **Steppe Buzzard**, **Booted Eagle** and most harriers. Likely rarities include **Goosander**, **Caspian Tern**, **Blue-cheeked Bee-eater** and **Pied Kingfisher**.

Nth.6. Agios Loukas Freshwater Lake North (Ayluga North) 35°08′08.1″N 33°54′42.6″E (35.135583, 33.911833) Larger than Freshwater Lake South, on the opposite side of the Famagusta–Nicosia (Gazimağusa–Lefkoşa) highway. Birds are difficult to see through the high reeds, so the lake is best viewed from the east, on the road to the Eastern Mediterranean University. This shallow lake attracts hundreds of waterbirds and general activity is impressive, because it is usually undisturbed. Regular species include **Gadwall**, **Eurasian Wigeon**, **Common Pintail**, **Garganey**, **Red-crested Pochard**, **Greater Flamingo**, **Little Bittern**, **Eurasian Spoonbill**, **Glossy Ibis**, **Purple Heron**, **Great** and **Cattle Egrets**, **Black-crowned Night Heron**, **Marsh Harrier**, **Little Crake**, **Marsh Sandpiper**, **Black-tailed Godwit**, **Black-winged Stilt** (breeds), **Whiskered** and **White-winged Terns**. Interesting sightings include **Common Crane**, **Bonelli's** and **Booted Eagles**.

Nth.7. Karaolos (Neapolis) Pond (Gülseren Pond) 35°08′23.7″N 33°55′04.6″E (35.139917, 33.917944) A medium-sized brackish lake with reedbeds and shallows, and a fairly stable water supply, located in the northern suburbs of Famagusta (Gazimağusa). Drive a short distance north on the Karpasia (Karpaz) road from the mosque roundabout ('Toros Centre') and turn right on any of the side roads after 1.2km, which all lead to the pond. Usually attracts large numbers of birds, including **Greater Flamingo**, loafing gulls and an interesting selection of waterfowl, herons and waders. **Moustached Warbler** and **Penduline Tit** are present in the reeds in winter.

Nth.8. Klapsides (Glapsides) Beach 35°09′36.8″N 33°54′42.0″E (35.160222, 33.911667) Best Aug–Mar. Located c.4km north of Famagusta, this dry river delta retains some seasonal brackish pools. At the south end of the beach, a small headland often hosts **Sandwich Tern**, **Audouin's Gull** and many other gulls in winter. Silver Beach (or Klapsides North) 35°10′00.3″N 33°54′21.0″E (35.166750, 33.905833) is just a few hundred metres north of Klapsides on another dry river delta next to the beach, where there are more seasonal brackish pools. At both sites there is a good selection of waders, including **Grey** and **Eurasian Golden Plovers**, **Greater Sand Plover**, **Pied Avocet**, **Spotted Redshank**, **Ruddy Turnstone**, **Eurasian Curlew** and many others. Rarities have included **Mute Swan**, **Lesser Spotted Eagle** and **Bar-tailed Godwit**.

Kouklia (Kukla) Wetlands is probably the country's largest seasonal wetland and is a great place to spend several hours birdwatching (Colin Richardson).

Nth.9. Kouklia (Köprülü) Dam (Kukla Wetlands) 35°07'09.4"N 33°45'55.0"E (35.119278, 33.765278) Best Sep–May. After winter rains, this large seasonal shallow dam is probably the island's premier wetland. Located at the south edge of the Mesaoria Plain, it is reached from the village of Kouklia (Köprülü), 2km north along a track from a turn-off near the mosque at the east end of the village. The creation of the Kukla Wetlands Birdwatching Centre (kuklawetlands.com) was funded by the EU and local council, and includes two well-sited bird hides. After a good wet season, the dam hosts hundreds of waterfowl, herons and waders, with notably large numbers of **Glossy Ibis**, **Ferruginous Duck** and **Ruff**. During migration, the dam and surrounding habitat attract a good selection of birds, including many raptors. Species include **Black-necked Grebe**, **Greater Flamingo**, **Great White Pelican**, **Eurasian Bittern**, **Red-crested Pochard**, **Bonelli's Eagle**, **Pallid Harrier**, **Demoiselle** and **Common Cranes**, **Broad-billed Sandpiper**, **Red-necked Phalarope**, **Little Crake**, **Calandra Lark**, **Common Rock Thrush**, **Great Spotted Cuckoo**, **Bluethroat**, **Moustached Warbler** and **Penduline Tit**. Large groups of **Collared Pratincole** often occur and breeding has been recorded. The country's first **Lesser Flamingo** was found here.

Nth.10. Galateia (Mehmetçik) Marsh 35°24'53.9"N 34°03'35.2"E (35.414971, 34.059789) A large, seasonal floodplain which can be dry for years. It depends on good winter and spring rains to become one of the top wetland sites. Located 1km south-west of Galateia (Mehmetçik), accessed from the main Bogazi–Rizokarpaso (Boğaz–Dipkarpaz) road to Karpasia. It attracts a selection of waterbirds, passerines and birds of prey. Some of the most interesting include **Great** and **Little Bitterns**, **Great Snipe**, **Little** and **Spotted Crakes** and **Common** and **Demoiselle Cranes**.

Nth.11. Karpasia Peninsula (Karpaz) 35°35'55.2"N 34°22'50.8"E (35.598672, 34.380765) Also known as the 'Panhandle', this extensive region, with Cape Apostolos Andreas (Zafer Burnu) at its easternmost point, occupies the north-east of the island. The area has unspoilt sandy beaches and a patchwork of rolling cereal fields and bush-covered hills, making it a magnet for migrant birds, especially in spring. Rizokarpaso (Dipkarpaz) to Cape Andreas is c.27km, but may need several hours to drive, stopping to look for interesting birds, with the last 15km the most productive. The road narrows on reaching the south coast of the peninsula

Karpasia, a beautiful area to visit in spring (Colin Richardson).

at the Blue Sea Hotel, and continues past more small hotels, tavernas and beach chalets, as the calls of **Black Francolin**, **Cyprus Wheatear**, and **Black-headed** and **Corn Buntings** become more frequent. **Little Owl** inhabits the abandoned buildings, **Cyprus Scops Owl** is widespread, and rocky promontories attract loafing **Audouin's Gulls**. The power lines that run the length of the road are ideal perches for rollers, bee-eaters, wheatears, shrikes, buntings and other migrants, and rarities like **White-breasted Kingfisher** and **Steppe Grey Shrike** have occurred. A good area is Golden Beach Valley (near the entrance to Burhan's Beach). Migrants are often found on bushes, fences and tracks, sometimes feeding in the extensive cereal fields. Recent rarities include **Blue-cheeked Bee-eater**, **Bimaculated Lark**, **Siberian Stonechat** and **Desert Finch**. There are likely to be harriers and other raptors too. Apostolos Andreas Monastery hosts a small colony of **Pallid Swifts**, and search the open areas and ground opposite the View Hotel for wagtails and wheatears. The road becomes a rough track for the last 4.5km to Cape Andreas, but with care it is suitable for most cars. About 1km before the cape turn left into a sheltered clearing surrounded by bushes. A line of telegraph poles and wires leads to a pumping station with dripping water. In addition to passerines, the area attracts many birds of prey, especially harriers and buzzards, and a **Bateleur** was found here in Apr 2007. Continue to the cape and on the right, just before the rocky hill, there is a grassy area which suits tired migrants. There have been several sightings of **Bar-tailed Lark**, **White-throated Robin**, **Rose-coloured Starling** and **Trumpeter Finch**. After another 300m there is a police post at the cape and the Klidhes Islands stretch out ahead. The nearest island hosts a colony of **Audouin's Gulls**, active Apr–Jul. Seawatching may produce shearwaters and other passing migrants.

Nth.12. Agios Filonas (Ayfilon) 35°37′47.7″N 34°22′26.5″E (35.629917, 34.374022) Return to Rizokarpaso and head north through the village on a winding route out of town following signs for the 'Oasis' (rooms and restaurant). Reaching the coast after 3.5km, turn right before the Oasis, following the coast road north-east for a few kilometres. The area is particularly good in spring for migrants which halt their passage before crossing the sea. The islet of Lefkonisos, 5km east of the Oasis, sometimes has breeding **Audouin's Gull**, and shearwaters are regular offshore. **Black-bellied Sandgrouse** has been seen several times.

Nth.13. Eptakomi/Platanissos (Yedikonuk/Balalan) Cliffs 35°28'15.6"N 34°02'24.5"E (35.471012, 34.040139) A short section of impressive sea cliffs on the north Karpasia coast 18km south-west of Gialousa (Yeni Erenköy) and 70km east of Kyrenia (Girne). The coast road provides excellent view of the cliffs. **Peregrine Falcon**, **Northern Raven** and a large colony of **Alpine Swift** nests here, while **Blue Rock Thrush** is usually present in winter. **Audouin's Gull** and shearwaters can be easily viewed passing offshore.

Nth.14. Flamoudi Dam (Azganlidere/Mersinlik Reservoir) 35°24'29.6"N 33°50'39.8"E (35.408231, 33.844392) A small but fairly productive dam with a reedbed. Seasonal, best visited Sep–May. Take the side road from the new north-coast road 52km east of Kyrenia (6km west of Davlos/Kaplica) signed Flamoudi (Mersinlik). For the best views of the dam, stop after 500m and walk 200m west across the cereal fields. It has hosted a good selection of birds including **Black-necked Grebe**, **Eurasian Bittern**, **Ferruginous Duck**, **Marsh** and **Hen Harriers**, **Spotted Crake** and **Marsh Sandpiper**. **Black Francolin** and **Cyprus Scops Owl** are resident in the area.

Nth.15. Pentadaktylos Peak (Beşparmak/Five Finger Mountain) 35°17'06.2"N 33°28'01.4"E (35.285049, 33.46705) Travel 8.5km east of Kyrenia (Girne) to where the coast road turns off to the left, then follow the airport road uphill over the Kyrenia Mountains. After 6.8km (before the summit) turn left on a side road. This track passes a dog's home, then runs downhill for 250m on a rough track, through light woodland until you reach the south face of the jagged 'five fingers'. The stony hillsides and bushes host breeding **Cyprus Wheatear**, **Blue Rock Thrush**, **Cyprus Warbler** and **Northern Raven**, and are one of the few regular wintering sites of **Rock Bunting**.

Nth.16. Kiados Dam (Serdarli Reservoir) 35°15'00.9"N 33°34'40.1"E (35.250241, 33.577794) Quite a productive small dam with reedbeds, attracting mainly waterfowl, on the south side of the Kyrenia Mountains. Although water levels fall in drought conditions, it rarely dries out. Reached from the village of Kiados (Serdarli) by heading 2.4km south-west on a gravel track. Other rough tracks exist to the dam from the

Pentadaktylos Peak, in the Kyrenia Mountains, hosts breeding Cyprus Wheatear, Blue Rock Thrush, Cyprus Warbler and Northern Raven (Colin Richardson).

main Lefkoniko (Geçitkale) to airport road. **Ferruginous Duck** breeds and several other waterfowl species occur including **Red-crested Pochard**. The area is good for migrants, including **Cyprus** and **Spectacled Warblers**, **European Bee-eater** and **Great Spotted Cuckoo**.

Nth.17. Kapouti/Agia Eirini (Kalkanli/Akdeniz) Dam 35°15'18.4"N 32°59'10.8"E (35.255102, 32.986334) A productive medium-sized dam with reedbeds and adjacent farmland, attracting a good selection of birds. Water levels vary; best visited Sep–May. Turn towards the ODTU Campus at the roundabout 5km north of Morphou (Güzelyurt) on the Kyrenia highway. After 800m – before the university entrance – turn left onto a dirt track. Drive 3.9km (the first 1km is very poor), eventually passing the remnants of a municipal dump, then turn right onto a narrow sandy track, head for a grey corrugated hut and reach the area above the dam after 700m. The dam hosts a large population of wintering waterfowl including **Black-necked Grebe**, **Ferruginous Duck** and one of the largest flocks of **Common Pochard** on the island. **Garganey** are common in autumn and spring. Other species include **Spur-winged Lapwing**, **Bonelli's Eagle**, **Long-legged Buzzard**, **Little** and **Spotted Crakes**, **Whiskered Tern**, **Great Spotted Cuckoo** and a selection of herons, waders and passerines. **White-headed Duck**, **Red-crested Pochard** and **Eleonora's Falcon** are occasional.

Nth.18. Livera (Sadrazamköy) Tank 35°23'16.5"N 32°56'54.4"E (35.387926, 32.948454) The main interest is a concrete water tank and a spring, beneath a large Black Mulberry tree. Situated 40km west of Kyrenia (Girne), the site is just 700m after the village of Livera (Sadrazamköy) heading towards Cape Kormakitis. The tank is just after Golya café. Best to sit quietly in your car, wait and watch. The trees and tank attract many migrants, including flycatchers, warblers and finches, and boasts an amazing list of rarities, mainly in autumn, including **Red-breasted Flycatcher**, **Isabelline Shrike**, **Pale Rockfinch**, **Eurasian Crimson-winged Finch**, **Red-fronted Serin**, **Common Rosefinch** and **Rose-coloured Starling**.

Nth.19. Cape Kormakitis (Koruçam Burnu) 35°23'44.1"N 32°55'51.2"E (35.395579, 32.930888) From the Livera (Sadrazamköy) Tank, continue west on the graded track towards the cape, first to a line of trees which crosses the road after 2km. This area is a very good place to look for birds, particularly flycatchers, warblers and

Livera Tank and the Black Mulberry tree, a migrant hotspot on the track to Cape Kormakitis (Colin Richardson).

redstarts. **Black-bellied Sandgrouse** were found here in winter 2014/15, and other interesting birds have included **Finsch's Wheatear** and **Cinereous Bunting**. Shearwaters congregate offshore and **Audouin's Gull** is regular. In Aug–Oct groups of herons, ibis, waders, gulls and terns pass the cape, and southbound raptors, including harriers, buzzards and **Western Osprey** make landfall here after leaving Turkey.

The following sites are also interesting, and include seasonal dams.

Nth.20. **Vasileia (Karşıyaka/Güzelyalı) Headland** 35°22′02.3″N 33°06′49.6″E (35.367317, 33.113774) This flat promontory, 20km west of Kyrenia (Girne), is a good place to observe coastal migration in autumn. Reached by surfaced road and a graded track onto the headland. Herons, ducks and other waterbirds can be seen offshore heading west in Aug–Sep. However, regular movements of **Yelkouan** and **Scopoli's Shearwaters** are often rather distant.

Nth.21. **Klepini Dam East (Uzundere/Arapkoy Reservoir)** 35°19′02.7″N 33°26′41.9″E (35.317421, 33.444983) Located along a rough track 500m south of the main road, halfway along the Kyrenia–Agios Amvrosios (Girne–Esentepe) highway. Attracts most endemic and local specialties, plus a selection of waterfowl, herons and waders in winter and on migration.

Nth.22. **Sygkrasi Dam (Sınırüstü Reservoir)** 35°16′54.5″N 33°51′06.6″E (35.281806, 33.851833) Located on the Mesaoria Plain next to the Bogazi–Nicosia (Boğaz–Lefkoşa) highway and easily accessible by car. Surrounded by farmland, goat-grazed meadows and cereal fields, it attracts a good selection of migrant ducks, herons and waders, including **Greater Sand Plover** and **Great Snipe**. **Rock Sparrow** has occurred nearby most winters.

Nth.23. **Kormakitis Dam (Koruçam Reservoir)** 35°19′24.8″N 32°59′44.6″E (35.323543, 32.995719) Difficult access along a rough track, set in a deep valley 3km west of Kormakitis (Koruçam) village. Reedbeds surrounded by hillsides and mixed forest, the waterfowl are rarely disturbed. Warblers, shrikes and waterfowl including **Ferruginous Duck**. Dries out in drought years.

Nth.24. **Agios Ermolaos Dam (Şirinevler/Polatdere Reservoir)** 35°15′42.9″N 33°10′09.5″E (35.261912, 33.169296) Easy access by car, adjacent to a good road in the southern Kyrenia foothills. Shallow dam with reedbeds, surrounded by mixed habitat. Waterfowl, migrant herons, harriers, swallows, swifts, wheatears and warblers. Dries out in drought years.

Female Cyprus Wheatear *Oenanthe cypriaca* at Anarita Park, near Paphos (Kate Bloomer).

Greylag Goose *Anser anser* wv

L 83. W 164. Rare, mainly in winter. Formerly common, now less than annual, late Oct–Apr, mainly late Dec–mid Feb. Usually in ones and twos, rarely flocks; occasionally with Greater White-fronted Geese. Escapes also reported. **Habitat** Salt lakes, marshes, dams, water treatment plants and cereal fields. **ID** Pink-billed birds, presumably eastern *A. a. rubrirostris* ('Eastern Greylag Goose') occur. Large goose with thick neck, ***heavy pink bill, pale greyish head and neck, and pink legs***. In flight, ***pale grey forewing***. Juvenile like adult but lacks sharply defined transverse lines on upperparts and dark belly marks. From other geese by combination of size, bill and leg colour, and strikingly pale forewing. **Voice** Similar to domestic goose; a loud, characteristic *ang-ang-ang* in flight.

Greater White-fronted Goose *Anser albifrons* wv

L 72. W 148. Uncommon winter visitor: late Oct–mid Mar, mostly late Nov–Feb. Rather local and declining. **Habitat** Salt lakes, marshes, dams, water treatment plants and cereal fields, mainly in south of island. **ID** The only regularly occurring large goose. Smaller than Greylag and warmer grey-brown, with slightly darker head and hindneck; ***orange legs*** and lacks silvery forewing. Wing pattern similar to Lesser White-fronted. Bill not heavy, but quite long, ***pink with white nail***; no eye-ring visible in field. ***Adult has large white area surrounding base of bill*** (rarely to forecrown) and ***black bars on underparts***. Juvenile browner, lacks white forehead and black bars on underparts; ***forehead and area round bill very dark, bill has dark nail***. **Voice** Musical and much higher pitched than other grey geese except Lesser White-fronted; repeated disyllabic, sometimes trisyllabic, with metallic, laughing quality *kow-lyow* or *lyo-lyck*.

Lesser White-fronted Goose *Anser erythropus* v

L 59. W 127. Vagrant, mainly in winter. **Habitat** Salt lakes, marshes, dams, water treatment plants and cereal fields. **ID** Resembles Greater White-fronted Goose, but smaller (a few overlap in size), with proportionately ***shorter neck and bill, steep forehead, yellow eye-ring*** and slightly longer wings, often extending beyond tail. Adult has ***extensive white on forehead to above eye***; upperparts dusky-brown with only dull transverse lines; underparts with only a few black blotches. Juvenile similar to juvenile Greater White-fronted Goose, but size, head and neck proportions, and yellow eye-ring separate. Compared to Greater White-fronted, has faster wingbeats, more compact silhouette and noticeably faster walk. **Voice** Higher pitched, squeakier and faster than Greater White-fronted; a di- or trisyllabic yelp *kow-yow, kyu-yu-yu* or piping *yi-yi-yi*.

Red-breasted Goose *Branta ruficollis* v

L 55. W 125. Vagrant. Occurs Nov–Feb. **Habitat** Favours salt lakes, marshes, dams and water treatment plants. **ID** Small, compact, dark goose with ***short, black bill***. Unmistakable pattern of ***black body, chestnut cheeks and breast*** (look dark at distance) ***bordered by white***; broad, shaggy, white line along flanks and white patch in front of eye. Juvenile has smaller chestnut cheek patch. Often mixes with the so-called 'grey geese'. Flight fast and agile.

Egyptian Goose *Alopochen aegyptiaca* v

L 68. W 144. Historically a scarce winter visitor, but very few records since 1960s. **Habitat** Fresh and brackish pools, marshes and dams. **ID** Slightly larger than Ruddy Shelduck, with broader wings, but similar wing pattern in flight. Otherwise ***brownish-buff*** with brownish or greyish upperparts, pale foreneck and face, and ***prominent dark eye patch***; dark collar and belly patch. Juvenile/immature browner, lacking eye patch, collar and belly patch, and with yellowish-grey legs. **Voice** Deep nasal braying.

ad
ad
Greylag Goose
juv
Greater White-fronted Goose
ad
ad
juv
Lesser White-fronted Goose
ad
Red-breasted Goose
ad
juv
Egyptian Goose

Mute Swan *Cygnus olor* wv

L 153. W 223. Scarce and irregular winter visitor, less than annual: late Nov–Apr, mainly late Dec–early Mar. Escapes of unknown origin also occur. **Habitat** Salt lakes and associated wetlands, occasionally coastal. **ID** *Orange bill with black knob* in adult (larger in male). *Typically, graceful S-curve to long neck, bill often pointing down*. Juvenile dingy brown with *black-based grey bill* gradually becoming pink then orange. Plumage grows increasingly white during first winter and spring, sometimes still partly brownish above until second winter. *In flight, wingbeats produce loud, rhythmic, singing sound vaou-vaou-vaou*. **Voice** Mostly silent.

Whooper Swan *Cygnus cygnus* v

L 153. W 231. Vagrant; winter. **Habitat** Salt lakes and associated wetlands. **ID** Size of Mute Swan with proportionately longer neck and bill; *latter black with prominent yellow base extending in wedge to below nostril or beyond*. Flight silent but with powerful wingbeats like Mute Swan. *Juvenile/immature similar to Bewick's, but note longer neck and wedge-shaped head profile*. Young often with parents during first winter. **Voice** Similar to Bewick's but deeper and stronger, with musical, trumpet-like quality *ahng-ha* or *ko-ko-ko*; often trisyllabic.

Bewick's Swan *Cygnus columbianus bewickii* v

L 121. W 195. Vagrant; winter, most records Jan. **Habitat** Salt lakes, freshwater pools and associated wetlands, occasionally coastal. **ID** *Smallest swan, with shorter neck and bill* than Mute and Whooper Swans. Similar to larger Whooper (both having straight neck), but *yellow on bill limited to base, ending well behind nostril*. Immature greyer than same-age Mute Swan, and *pink bill lacks black base*, becoming white and later yellow during first winter. From immature Whooper by *shorter neck and bill, and more rounded head*. In flight, wingbeats slightly faster, more goose-like, with no humming sound. **Voice** Similar to Whooper but higher pitched, monosyllabic or disyllabic (Whooper often has three syllables), sometimes recalling distant barking dogs.

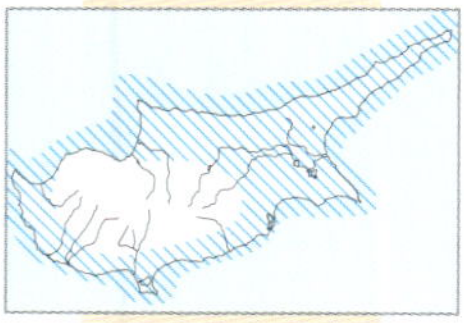

Great White Pelican *Pelecanus onocrotalus* pm

L 140–175. W 270–330. Uncommon passage migrant. Mainly autumn, peaking Aug–Nov, but records all months, sometimes in large flocks. Stranded groups may fly considerable distances, visiting suitable lakes and dams across the island. **Habitat** Freshwater dams and pools, occasionally salt lakes, even small marshes or ornamental ponds if food is available. **ID** Large with huge wingspan. Adult *white with contrasting, solid black flight feathers below*, but body tinged yellowish-rosy in breeding plumage (Dalmatian Pelican has greyish underwing, and body appears greyish-white). At close range note *dark eye surrounded by naked rosy skin, fleshy-yellow legs and pointed forehead feathers where they meet culmen*; latter characters are also useful to separate immature from similar Dalmatian Pelican. Immature Great White Pelican has *clearly darker grey-brown upperparts than the grey-buff Dalmatian*. Flight consists of a few slow wingbeats followed by a glide; flocks often fly in regular lines, or circle in formation.

Dalmatian Pelican *Pelecanus crispus* v

L 150. W 310–345. Vagrant. Mainly late autumn and early spring, very rarely in winter. **Habitat** Salt lakes, freshwater pools and associated wetlands. **ID** Resembles Great White Pelican but separated in flight by *greyish underwing with pale band in centre and greyish-white body* (in Great White, flight feathers below solidly black, and white body tinged yellowish-rosy). At close range, *nape feathers curl upwards* (drooping in Great White), *pale eye* (dark in Great White) *and grey legs*; also shape of bare skin around eye and of feathers where they meet culmen useful at all ages. Immature dirty white below, *pale grey-buff above* (same-age Great White is dark grey-brown above).

Mute Swan
imm
ad
juv
Whooper Swan
ad
juv
Bewick's Swan
ad
juv
ad winter
Great White Pelican
ad winter
juv
Dalmatian Pelican

PLATE 3: DABBLING DUCKS

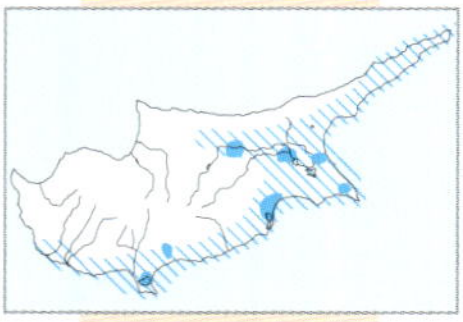

Common Shelduck *Tadorna tadorna* WV, pm

L 63. W 122. Common winter visitor, uncommon passage migrant. Occurs Oct–early May, mainly late Nov–Apr, peaking late Dec–early Mar. Occasionally in summer. **Habitat** Mostly salt lakes; also dams, water treatment plants and marshes. **ID** Large with white body, ***black head with green gloss, broad chestnut band on breast and prominent red bill with fleshy knob in breeding male***. In flight, black flight feathers contrast with white upper- and underwing-coverts and black shoulders. Juvenile lacks chestnut band, has grey-brown head, hindneck and upperparts, and pale greyish bill. **Voice** Not often heard; male has high whistling *siss-siss-siss*, female a whinnying *gehe-hehe-heh-eh*; sometimes a nasal *ah-hang*.

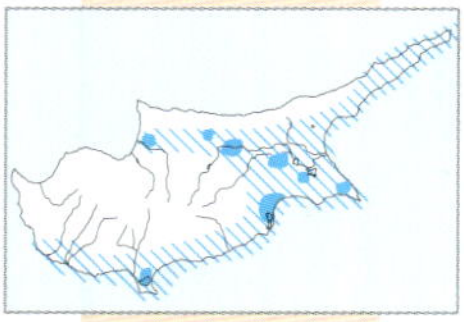

Ruddy Shelduck *Tadorna ferruginea* wv

L 64. W 133. Uncommon winter visitor. Less than annual. Occurs Sep–Apr, mainly Dec–Feb. Some may also occur on passage. **Habitat** Salt lakes, dams, water treatment plants and marshes. **ID** Size of Common Shelduck with ***orange-chestnut body*** and paler cinnamon-buff head; ***black flight feathers*** with green speculum, ***striking white forewing and underwing-coverts*** (similar to Egyptian Goose). Female and non-breeding male lack black collar; female also has paler head. Juvenile like adult female, but back browner. From Egyptian Goose by shape, narrower more pointed wings, head pattern and darker chestnut underparts. **Voice** Not often heard. Nasal, trumpeting and rather penetrating *ang, ang*.

Gadwall *Mareca strepera* pm, wv, ob?

L 51. W 89. Uncommon passage migrant and winter visitor. One of the least common regular ducks. Occurs mid Sep–mid Apr, mainly late Dec–Feb. Sometimes remains to late spring and may breed occasionally. **Habitat** Any wetland. **ID** Medium-sized, slightly built dabbling duck. In flight, rapid wingbeats and pointed wings like Eurasian Wigeon. ***Male mainly dark grey with black around tail; in flight, white speculum bordered black in front, and white belly***. Female and juvenile resemble larger and heavier female Mallard, but tail grey-brown (not white), and thin dark bill is orange on sides; ***in flight, shows white speculum near body*** (often smaller in male) and ***white belly*** bordered by dark flanks. **Voice** Male has a rasping, low *rrep* call. Female's call resembles Mallard's but is higher pitched.

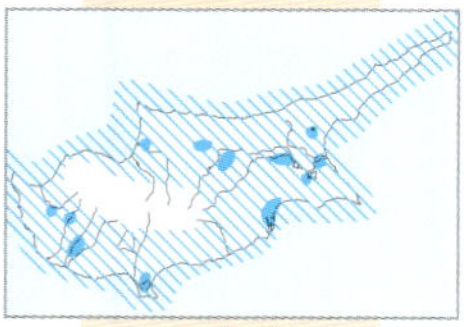

Eurasian Wigeon *Mareca penelope* wv, pm

L 48. W 80. Uncommon winter visitor and passage migrant. Occurs late Sep–early May, mainly late Nov–Feb. Passage Mar–Apr. Has declined in recent years, especially in south. **Habitat** Marshes and lakes with suitable grazing. **ID** ***Rather rufous duck*** with short neck, rounded, ***steep forehead and small blue-grey bill. In all plumages has white belly patch***, obvious in flight and when grazing. Adult ***male has creamy-yellow forehead***. Adult female rufous to greyish-brown; lacks white forewing. Juvenile resembles adult female (immature male does not usually assume white forewing until second winter). In flight, in all plumages, has narrow wings, pointed tail and contrasting white belly patch; ***white forewing contrasting with blackish-green speculum is distinctive in adult male***. Female and juvenile have grey-brown forewing, darker flight feathers and black speculum. **Voice** Male gives characteristic clear whistle *wheeooo*, both in flight and on water; female growls in flight.

Mallard *Anas platyrhynchos* RB, PM, WV

L 56. W 95. Common breeding resident, passage migrant and winter visitor. Migrants occur from Aug, with peak numbers mid Oct–Jan, numbers decline quickly as birds depart from late Feb. **Habitat** Breeds mainly on fresh water, plus salt lakes and water treatment plants in winter. **ID** Large dabbling duck. Male has ***grey wings, dark blue speculum distinctly bordered white, and white underwing-coverts***. Female mottled brown like other female ducks, from which separated by size, bill shape and colour, and wing pattern. Male in eclipse resembles female, but bill greenish-yellow. **Voice** Male has a soft nasal *raehb* and, during courtship, a weak, high-pitched whistle *piu*. Female gives familiar deep quacking.

♀
♀
♂
♂
♀
Ruddy Shelduck
juv
Common Shelduck
♀
♀
♂
Gadwall
♀
♂
Eurasian Wigeon
♀
♀
♂
Mallard

PLATE 4: DABBLING AND DIVING DUCKS

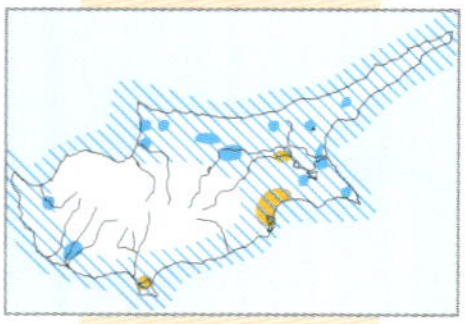

Northern Shoveler *Spatula clypeata* — WV, PM, ob

L 51. W 78. Very common winter visitor, common passage migrant and occasional breeder. Migrants occur from mid Aug or earlier, to May; winter peak late Dec–mid Jan; peak spring passage mid Mar and most depart by mid May. **Habitat** Water treatment plants, fresh and saltwater lakes and marshes. **ID** ***Huge spatulate bill***. Swims with front end low and bill often dabbling in water. In flight, wings appear set far back. Male unmistakable; in flight, looks black, white and chestnut with ***distinctive blue forewing***. Female and juvenile on water resemble female Mallard and Gadwall, but bill distinctive; in flight, shows bluish forewing, somewhat resembling smaller male Garganey. Male in eclipse largely resembles female, but brighter blue forewing. **Voice** Male call a hollow, double-note *g-dunk, g-dunk*, often in flight, while female simultaneously quacks *pe-ett*.

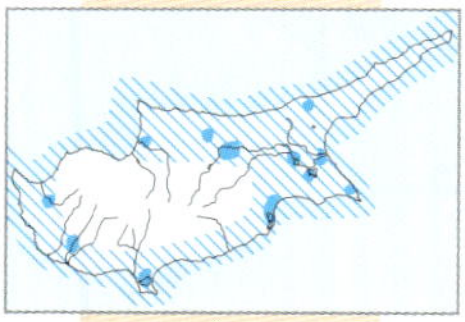

Northern Pintail *Anas acuta* — WV, PM

L 56 (excluding long tail feathers of male). W 88. Fairly common passage migrant and winter visitor. Vanguard from late Jul, mainly mid Sep–early May, occasionally in summer. Spring peak passage late Feb–early Apr often coastal, peaking mid Mar; autumn peak late Oct–early Nov. **Habitat** Shallow inland waters. **ID** Slim and elegant dabbling duck. Male has ***white neck and underparts with dark head and tail, and dark slender bill***. Female recalls other female dabbling ducks, but note ***dark bill, greyer plumage and much slimmer appearance***. In flight both sexes show distinctly ***longer neck and tail***, and slender, more pointed wings than other ducks; also note ***green speculum with conspicuous white border at rear***. Female has distinctive ***white border at rear of secondaries***; upperwing otherwise brownish. **Voice** Male a low-pitched weak whistle; female a hoarse quack.

Garganey *Spatula querquedula* — PM, ob

L 39. W 63. Very common passage migrant and occasional breeder. Spring passage late Feb–mid May, peaking mid Mar–Apr. Autumn passage Jul–mid Nov, peaking late Aug–mid Sep. Breeds Apr–Jun some years. **Habitat** Mainly freshwater wetlands; large numbers on coastal passage. **ID** Small dabbling duck. ***Male shows long white stripe on head and, in flight, striking blue-grey forewing***. Female similar to slighter female Eurasian Teal, but ***longer, heavier bill and more contrasting dark and pale head-stripes, widening into pale patch at bill base***, and whiter throat. In flight, female's forewing is slightly paler and white border along secondaries is distinctive (Eurasian Teal has white wingbar on mid-wing in front of speculum). Male in eclipse is like female, but wing pattern as adult male. Juvenile similar to female. **Voice** Male call a dry, drawn-out rattled *knerreck*; female a short, sharp quack.

Eurasian Teal *Anas crecca* — WV, PM

L 36. W 61. Very common winter visitor and passage migrant. Vanguard from late Jul, mainly Aug–mid Apr, peaking mid Oct–mid Mar; a few all year. **Habitat** Most wetlands, including salt lakes, marshes and water treatment plants. **ID** Commonest small duck; readily takes flight, both sexes showing white belly and ***greyish underwing with light band in centre***. Adult male distinctive; female rather similar to Garganey (which see for differences). Often in fast-flying, tightly knit, twisting and turning flocks. **Voice** Male has far-carrying ringing whistle *kreek-kreek*; female a high-pitched, nasal quacking.

Marbled Duck *Marmaronetta angustirostris* — v, FB

L 41. W 65. Vagrant. Mainly spring and summer months. Bred until early 20th century when fairly common. **Habitat** Salt lakes and similar wetlands. **ID** In flight resembles small female Northern Pintail, with fairly long neck, wings and tail. On water identified by ***pale plumage, dappled dark and cream*** (marbled) with ***dark oval patch around eye***. Head large and rounded with steep forehead and, in adult, bulky crest on lower nape. In flight, wings fairly pale with even slightly paler secondaries but ***no speculum***. Often secretive, hiding in vegetation. **Voice** Low nasal wheezing *jeak* or double whistles.

♀
♂
Northern Shoveler
♀
♂
♀
Northern Pintail
♀
♂
Garganey
♀
♂
Eurasian Teal
ad
Marbled Duck

Red-crested Pochard *Netta rufina* wv, pm, ob

L 56. W 87. Uncommon winter visitor and passage migrant in variable numbers, and occasional breeder. Occurs mid Nov–early Mar, mainly Jan–Feb. Breeds Mar–Jun, with family parties remaining until mid Aug. **Habitat** Freshwater lakes and large pools, water treatment plants and marshes; rarely salt lakes. **ID** Large diving duck which sits high on water. Male unmistakable *with large dark orange head and red bill*; in eclipse resembles female but bill all red. Female has pale grey cheeks contrasting with dark crown; dark bill has pink band near tip. In flight shows broad white wingbar. **Voice** Male has a double *weep-weep*; female a grating *keerr*.

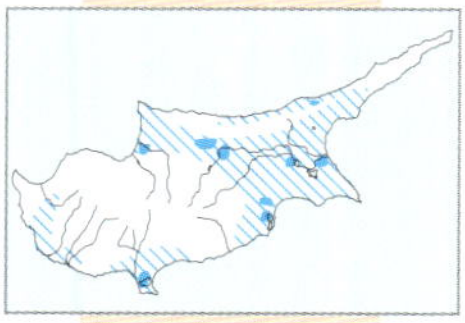

Common Pochard *Aythya ferina* WV, pm

L 45. W 80. Fairly common winter visitor and passage migrant. Occurs mid Sep–Apr, mainly late Nov–Feb, peaking Jan. Spring departure Feb–mid Mar, occasionally later. **Habitat** Freshwater lakes and large pools, water treatment plants and marshes, mainly northern Cyprus. **ID** *Sloping forehead grading into long bill* is characteristic. Male has *chestnut head and neck, contrasting with black breast*; black bill has pale blue-grey band in centre. In eclipse resembles female but greyer above. Female has dull brownish head and breast with paler chin and eyestripe; dark bill becomes paler towards broad black tip. In flight looks longer and plumper than Tufted Duck; greyish-brown wings (paler in male) show *indistinct pale grey wingbar*. **Voice** Courting male has low whistle; also hoarse wheezing note; female has rough harsh *krra-krra*.

Ferruginous Duck *Aythya nyroca* WV, pm, rb

L 40. W 66. Fairly common winter visitor, passage migrant and breeding resident. Migrants occur mid Oct–mid Mar, peaking mid Dec; breeders increasing and spreading at suitable wetlands. **Habitat** Well-vegetated freshwater pools, dams and reedbeds, especially when breeding. **ID** Slightly smaller than Tufted Duck, head shape close to Common Pochard. *Compared to Tufted Duck has higher crown, flatter sloping forehead and longer, dark grey bill with black nail. Male rich chestnut-brown with white eye and sharply defined white undertail-coverts*. Female and immature have dark eyes (white in one-year-old male), and lack whitish at base of bill of some Tufted Ducks; *female has warmer brown head than Tufted* (which has yellow eye); pure white and sharply defined white undertail-coverts (sometimes seen in Tufted). *First-years separated by head shape, length and pattern of bill, and dark eye*. In flight, *broader, more conspicuous white wingbar than Tufted Duck*. **Voice** In flight, high-pitched *crr-err*.

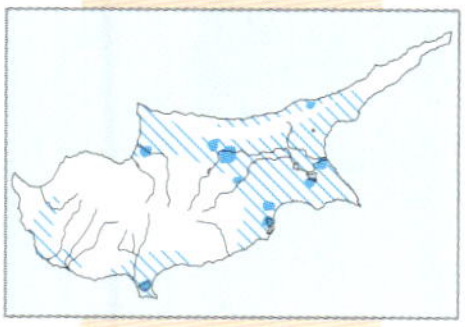

Tufted Duck *Aythya fuligula* wv, pm

L 42. W 70. Scarce winter visitor and passage migrant. Occurs mid Sep–Apr, mainly mid Nov–Mar. **Habitat** Freshwater lakes, ponds and dams. **ID** Small diving duck with roundish head but fairly steep forehead; *crest on nape*, long and drooping in male, tiny in female. *Bill blue-grey* (darker in female) *with broad black tip. Male has black upperparts and purple head sheen*. Female and male in eclipse have brownish sides to body; female sometimes a whitish band at base of bill. In summer, female has *darkish back, short crest and broad black tip to bill*. Some females show white undertail-coverts. In rapid flight has conspicuous white wingbar, almost full length of wing.

Greater Scaup *Aythya marila* v

L 46. W 79. Vagrant. Mainly Nov–Jan. **Habitat** Freshwater pools, dams and other inland waters. **ID** Larger than Tufted Duck with larger, more rounded head, *no crest and black confined to nail of longer blue-grey bill*. Male has *pale grey back and greenish sheen to head*. Female and eclipse male have *dark back washed greyish; bold white band at base of bill* (sometimes approached by female Tufted); in spring also a *pale patch on ear-coverts*, absent or indistinct in winter. Some first-winters lack white at base of bill, but separated at close range by shape of head, bill pattern and some greyish mottling on back.

Red-crested Pochard

Common Pochard

Ferruginous Duck

Tufted Duck

Greater Scaup

Common Scoter *Melanitta nigra* v

L 49. W 84. Vagrant, one very old Mar record. **Habitat** Mainly maritime. **ID** Medium-sized dark sea duck. ***Male all black with yellow/orange on upper mandible and black wings in flight***, though flight feathers slightly paler. Female has dark brown body and crown with ***contrasting greyish-buff cheeks*** and dark wings in flight. From Velvet Scoter *Melanitta fusca* (not recorded in Cyprus, but potential vagrant) by head and wing pattern.

Common Goldeneye *Bucephala clangula* v

L 46. W 73. Vagrant. Mainly Dec–Apr. **Habitat** Freshwater pools, lakes and marshes. **ID** Fairly small, easily identified sea duck with ***large rounded head***, yellow eye (adult) and much white in wing, obvious in flight and often on folded wing too. Male largely white with black back and rump, and black head, glossed green; ***large white circular patch between eye and bill***. Female greyish with ***contrasting brown head and white belly***. In fast flight easily recognised by round head, short neck, ***nearly white inner wing*** and dark underwing; rapid wingbeats, male producing characteristic ***whistling wing noise***, audible at long range.

Smew *Mergus albellus* v

L 41. W 62. Vagrant. Mainly Dec–Feb. **Habitat** Inland wetlands. **ID** Small sawbill duck. ***Male largely white with black mask and black back***. Female resembles female Common Goldeneye, but ***white cheeks distinctive***; from female Red-crested Pochard by size and wing pattern; much paler than female Common Scoter. Male in eclipse (resembles female) until Nov. Flight fast and agile, and shape slender; wings black in male with white patch on forewing; female similar, but has grey-white forewing and brown flight feathers.

Ruddy Duck *Oxyura jamaicensis* v?, E/I?

L 40. W 60. Vagrant or escape. Accidental visitor from unknown established populations. **Habitat** Usually shallow lakes. **ID** A stifftail duck, similar to White-headed, but in all plumages lacks a swollen base to upper mandible. Male Ruddy Duck always differs from White-headed in having a noticeably dark chestnut crown and hindneck. Female and juvenile have less contrasting, more diffuse, face pattern compared to White-headed on buffish, not white, cheeks.

White-headed Duck *Oxyura leucocephala* wv, pm?

L 46. W 66. Scarce winter visitor, rare in autumn. Occurs Nov–early Mar, mainly Dec–early Feb. Usually alone, occasionally small groups. **Habitat** Reedy, shallow lakes and brackish lagoons. **ID** Rather small, strange-looking diving duck. Male unmistakable by ***white head, large blue bill and long stiff tail often held cocked***. Female buffish-brown with dark brown head and variable ***grey- and dark-striped cheeks***; bill grey, swollen at base. Male in eclipse more female-like, but usually has white cheeks. Juvenile duller and paler than female. In flight looks elongated with short uniform wings set far forward on body.

Common Scoter
Common Goldeneye
Smew
Ruddy Duck
White-headed Duck

Red-breasted Merganser *Mergus serrator* wv

L 56. W 78. Scarce winter visitor, less than annual. Occurs late Nov–mid Mar. **Habitat** Coastal inshore waters, occasionally brackish pools. **ID** *Conspicuous ragged wispy crest in both sexes*. In fast direct flight, appears elongated with long head and neck, and shows much white on inner wing. *Male's greenish-black head, thin red bill, white collar and rusty black-spotted breast* easily separate it. Female and juvenile closely recall female Goosander, but smaller and slimmer-billed, with more brownish (less greyish) upperparts, *less contrasting head pattern, and no sharp demarcation between brown head and greyish-buff neck and breast*, flanks darkish grey.

Goosander *Mergus merganser* v

L 65. W 90. Vagrant. Mainly winter and early spring. **Habitat** Lakes and reservoirs. **ID** Large and elongated both on water and in flight. Flight strong, fast and direct, sometimes high over land. Nearly all-white inner wing very striking in flight; when overhead looks white with black head. Female and juvenile easily confused with female Red-breasted Merganser, but *chestnut-brown head is sharply demarcated from white chin and whitish neck, and has floppy crest*; in flight, larger and paler upperparts with more white on inner wing than female Red-breasted Merganser; flanks pale greyish. [Alt: Common Merganser.]

Chukar Partridge *Alectoris chukar* RB, E/I

L 33. Resident and generally sedentary. Rather skulking Feb–early Apr when nesting. Many introduced in autumn for hunting. **Habitat** Occurs all altitudes in many habitats, including farmland, salt scrub, olive groves, clearings and lightly wooded hillsides. Often in groups of 10–20 (cf. Black Francolin, which is usually solitary or in pairs). **ID** A distinctive partridge with *conspicuous black-and-white head pattern and barred flanks. Characteristic voice is often first indication of its presence*. Wary, runs fast; upright posture when alert. **Voice** Accelerating *ka-ka-kaka-kaka-kaka*... followed by a chuckling *chukara-chukar-chukara-chukar* frequently heard. *A. c. cypriotes* is a very common breeding resident, and this race also occurs on the Greek islands and in Turkey.

Black Francolin *Francolinus francolinus* RB

L 35. W 53. Common breeding resident. Very loud and conspicuous Mar–May. Usually in singles or pairs, rarely more (cf. Chukar). Breeds Mar–early Jun, adults with young May–Aug. **Habitat** Irrigated farmland, meadows and cereal fields in low-lying coastal areas and foothills. **ID** Large, plump and stub-tailed. More often heard than seen and hard to flush. Male distinctive, *mostly black with chestnut collar*, white spots on mantle and breast-sides merge into *white chevrons on flanks; obvious white patch behind eye*. In flight, black- and brown-barred wings and blackish tail. Female *warm brown* with paler head and *arrowhead-shaped feather centres*, most obvious on lower breast and flanks; *indistinct chestnut patch on hindneck*. In flight note dark tail with blackish outer feathers. **Voice** Male calls most intensely at sunrise and evening, reaching peak early–mid April, *loud, penetrating* and very distinctive with *grating* character, mostly from a mound or a bush *gldlri – djjii – djji-djji-djjii – djji-djjii*, first note lower and second and fifth stressed. Less vocal Aug–Dec.

Common Quail *Coturnix coturnix* PM, wv, mb

L 17. W 33. Common passage migrant, scarce winter visitor and uncommon breeding visitor. Recorded all months. Passage from Mar, peaking late Mar–Apr, with small numbers until early Jun and breeders later. Fewer Sep–mid Nov, although 'thousands' reportedly trapped by hunters in autumn. **Habitat** Cereal crops, meadows and rough grassland. **ID** Small gamebird, *more often heard than seen*, and hard to flush. In flight, *size of Common Starling*; rather pointed and narrow wings held bowed, fast shallow wingbeats, *plain wings and striped back*. Creeps about inconspicuously on ground. Female has paler head pattern and lacks neck-band of male. **Voice** Characteristic sound in open country over much of Cyprus, especially in farmland, by day or night, an explosive, rhythmically repeated *trisyllabic whistle pit, pil-it* (rendered as 'wet my lips').

Red-breasted Merganser

Goosander

Chukar Partridge

Black Francolin

Common Quail

European Storm Petrel *Hydrobates pelagicus* V

L 15. **W** 38. Vagrant. Rare east Mediterranean; *H. p. melitensis*, considered by some a separate species 'Mediterranean Storm Petrel'. **Habitat** Maritime, may follow ships. **ID** Small with square-ended tail, confusable with Wilson's Storm Petrel; distinguished by ***white stripe on underwing***, shorter wings, fairly uniform dark upperwing and ***less extensive white rump. Feet do not project beyond tail. Flight weak with nearly continuously fluttering wingbeats*** interspersed by short glides. Differs from larger Leach's Storm Petrel in shorter wings, shape of rump patch, darker upperwing, white on underwing-coverts, and flight.

Leach's Storm Petrel *Oceanodroma leucorhoa* No records

L 20. **W** 46. Very rare winter visitor to east Mediterranean, and potentially to Cyprus, from North Atlantic. **Habitat** Maritime. **ID** Sooty-brown, often looking black, with a pale bar across upperwing-coverts and rather ***dirty white rump with a dark central band (noticeable only at close range). Flight erratic and bounding, at times recalling a Black Tern***, with long, rather angled wings and deep wingbeats. Smaller European Storm Petrel lacks noticeable covert bar, has square tail, ***white on underwing*** and different flight.

Scopoli's Shearwater *Calonectris diomedea* PM

L 46. **W** 113. Fairly common migrant or wanderer, probably dispersing from colonies in Greece and Malta. Passage late Mar–early Jun on south coast, peaking early–mid April. In autumn, most occur late Jun–Oct, mainly on north coast. Usually fewer than ten birds together, but hundreds occur Oct. Virtually absent in winter. **Habitat** Maritime. **ID** Large shearwater with ***yellowish bill and no clear demarcation between grey-brown head and white chin and throat***. Underwing white with narrow brownish margins and may show ill-defined whitish band at base of uppertail. Flight relaxed with fairly slow wingbeats interspersed by long glides, often in arcs above waves, on ***characteristic bowed wings***. Scopoli's Shearwater has recently been split from Cory's Shearwater (latter has *not* been reliably recorded in east Mediterranean) and they are difficult to separate. ***Scopoli's is slightly smaller than Cory's, with thinner, duller bill, slightly narrower wings with white on underwing typically extending as 'fingers' on primaries,*** but separation obviously only possible at close range or by careful examination of photographs.

Cory's Shearwater *Calonectris borealis* No records

L 46. **W** 113. Potential vagrant. Not reliably recorded in Middle East, but could occur (from North Atlantic). **Habitat** Maritime. **ID** Large shearwater almost identical to Scopoli's Shearwater (which see), but slightly ***heavier with brighter yellow bill and more solidly demarcated dark tips to primaries below***. Most will be indistinguishable unless seen well at very close range.

Yelkouan Shearwater *Puffinus yelkouan* PM

L 33. **W** 82. Fairly common migrant or wanderer, equally regular at both seasons. Main spring passage is west along south coast late Mar–early May, peaking mid Apr, while autumn passage is east along north coast Aug–early Nov, peaking Oct. **Habitat** Maritime. **ID** ***Noticeably smaller than Scopoli's Shearwater, with dark brown upperparts*** (can look black at a distance) and ***white underparts; in worn plumage can show smudgy brown on breast, flanks and undertail***. Underwing sullied white with flight feathers variably brownish. Rather fluttering flight low over water, interspersed by short glides.

ad
European Storm Petrel
ad
Leach's Storm Petrel
ad
Scopoli's Shearwater
ad
Cory's Shearwater
ad
Yelkouan Shearwater

PLATE 9: GREBES

Little Grebe *Tachybaptus ruficollis* WV, PM, RB

L 27. W 43. Common winter visitor, passage migrant and resident. Breeds mid Jan–Jun. Few spring migrants Mar–Apr. Noticeable autumn influx Aug–Dec combines with dispersing local juveniles. **Habitat** Well-vegetated freshwater pools and marshes, dams and water treatment plants in winter. **ID** Small size, *blunt-ended body*, short neck and bill. *In flight, amount of white on secondaries and inner primaries varies, being obvious in race* capensis (which occurs in most of region) but almost absent in *ruficollis* (western Turkey). Adult has *bright chestnut throat and cheeks, and conspicuous yellow gape patch*. In winter, adult and young paler brown above, buff below, variably mixed dull chestnut on foreneck (adult); pale gape often reduced or absent. Juvenile has white striped head. Dives with fast jump; when alarmed, dives rather than flies. **Voice** When breeding, more often heard than seen; distinctive *high-pitched trill*, recalling whinny of horse.

Red-necked Grebe *Podiceps grisegena* V

L 45. W 81. Vagrant. Mainly Jul–Dec, occasionally spring. **Habitat** Lakes and inshore coastal waters. **ID** Close in size and shape to Great Crested Grebe, but head more rounded and *neck shorter. Bill black with diagnostic yellow base (adult)* or yellow extends diffusely towards tip of lower mandible (juvenile/first-winter). *Black cap to eye level, pale greyish cheeks contrasting with dark crown and dusky neck*. From smaller grebes in winter plumage by size, long bill and oblong head. In flight, *white patch on secondaries and inner forewing smaller and not connected over body*; wing pattern thus like Horned Grebe, but more white at base of forewing.

Great Crested Grebe *Podiceps cristatus* wv, pm, ob

L 49. W 88. Uncommon winter visitor and passage migrant. Occurs Sep–mid May, mainly from mid Nov, most depart by mid Apr. May breed occasionally. **Habitat** Fresh and brackish lakes and pools, occasionally coastal waters. **ID** *Large grebe with long white slender neck and pink, dagger-like bill* held horizontally. In flight, extended neck and feet held below line of body; wingbeats rapid, *large white patch on secondaries, white border to forewing and on shoulders along body*. Easily visible on open water and usually submerges smoothly, without leaping. Winter plumage lacks tippets, but retains short, dark grey crest; *black lores and narrow white line over eye* make eye clearly visible at distance; juvenile has black-and-white striped head and neck.

Horned Grebe *Podiceps auritus* V

L 35. W 62. Vagrant. Mainly Nov–Jan, rarely in spring. **Habitat** Fresh and coastal waters. **ID** Similar in size to Black-necked Grebe; smaller and has relatively shorter bill than Red-necked Grebe. *Winter plumage strikingly black and white; white cheeks and foreneck contrast with black cap down to eye and black upperparts*. From Black-necked Grebe in winter plumage by *gleaming white foreneck, white cheeks and straight bill*. In flight, distinctly smaller, with more rapid wingbeats than Red-necked; upperwing shows white wing patch confined to secondaries (extends to inner primaries in Black-necked) and some white on inner forewing (unlike Black-necked). [Alt: Slavonian Grebe.]

Black-necked Grebe *Podiceps nigricollis* WV, PM, ob

L 31. W 58. Fairly common winter visitor and passage migrant. Occurs from Aug, mainly mid Nov–mid Mar, most departing by mid Apr. Some spring migration mid–late Feb. May breed occasionally. **Habitat** Fresh and brackish lakes and pools, very occasionally coastal waters. **ID** Small grebe with *short, uptilted bill, steep forecrown and peaked head*; sometimes looks puff-backed, recalling Little Grebe. Breeding plumage *mostly black (including foreneck)* with golden ear-coverts and chestnut flanks. Winter plumage basically black and white, recalling Horned or even Red-necked Grebes. From Red-necked by size, short uptilted bill, red eye, steep forehead and more rounded back. From Horned (with which most easily confused) by *head and bill shape, cheek pattern and less clean black-and-white appearance*. Uptilted bill can be difficult to see at distance. In flight, long white wing patch extends to inner primaries (unlike Horned); also lacks white on inner forewing (present in all other grebes). [Alt: Eared Grebe.]

Little Grebe

Red-necked Grebe

Great Crested Grebe

Horned Grebe

Black-necked Grebe

PLATE 10: FLAMINGOS, IBIS, SPOONBILL AND BITTERN

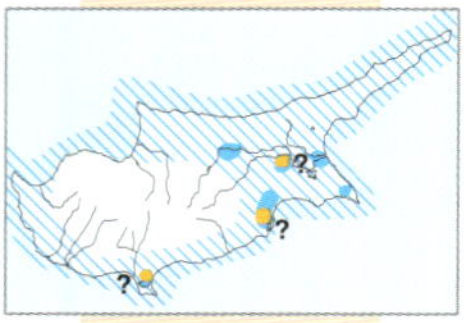

Greater Flamingo *Phoenicopterus roseus* PM, WV, ob?

L 130. W 155. Common passage migrant and winter visitor. Occurs all months, mainly mid Aug–Apr. Numbers arrive at salt lakes after heavy winter rains, some from Oct, peak Nov–Jan. Seen nest-building, including at Akrotiri and Larnaca. **Habitat** Salt lakes, brackish and shallow freshwater pools, marshes and sewage lagoons. **ID** Very large, long-legged, long-necked, pinkish wading bird with characteristic bill shape. White plumage of adults gradually acquires a pink hue and red wing-coverts. Juveniles, which are half the height of adults, are greyish with brown markings. Most readily separated from Lesser Flamingo by larger size (but beware size variation in Greater) and *pink (adult) or greyish (juvenile) bill with black tip* (all-blackish bill in Lesser Flamingo). **Voice** Cacophony of deep honks and grunts, strongly recalling domestic geese.

Lesser Flamingo *Phoeniconaias minor* v

L 85. W 130. Vagrant. **ID** *Smaller than Greater Flamingo and generally deeper and brighter pink with all-blackish bill,* although very close views reveal dark carmine near tip. In full adult plumage has deep rose-pink face bordering bill (lacking in Greater) and long scapulars (pinkish-white to white in Greater). In flight, shows rose-pink patch on central upperwing-coverts, bordered pinkish-white. Iris red (pale yellow in Greater). *Juvenile similar to Greater Flamingo in plumage but blackish bill readily identifies it.* Smaller size often difficult to establish in lone birds; also immatures of both species are smaller than adults. *In flight, note shorter neck and legs, faster wingbeats and more uptilted chin* than Greater Flamingo.

Glossy Ibis *Plegadis falcinellus* PM, mb

L 65. W 90. Common passage migrant and local breeding visitor. Occurs on passage Mar–May, peaking Apr; and late Jul–mid Oct, mainly Aug–early Oct. A recent coloniser, nesting Apr–Jun. Some present all year. **Habitat** Freshwater wetlands and marshes; nests colonially in trees and reeds. Mainly coastal on migration. **ID** *Blackish with long, decurved bill* but close views reveal adult to have *deep purple-chestnut plumage,* glossed green on wings. In breeding season has white marks at base of dull pink bill; in winter, bill brownish with fine pale streaks on head and neck. Juvenile much duller. Fast wingbeats in flight often interspersed by long glides; frequently flies in linear formation. **Voice** A guttoral *kraa kra* at colonies or when disturbed.

Eurasian Spoonbill *Platalea leucorodia* pm

L 85. W 120. Uncommon passage migrant. Occurs all months, mostly late Aug–mid Nov, peaking mid Sep–mid Oct. Less frequent mid Feb–early Jun, peaking mid–late Apr. Occasional in summer; rare in winter. **Habitat** Shallow open water and mudflats. **ID** Heron-sized with *all ivory-white plumage, characteristic black spatulate bill with yellow tip, and black legs.* Adult has nape plumes and yellowish neck-band, both lost in winter. Immature has dull flesh-coloured bill and legs, and black wingtips. In flight, neck extended and wingbeats fast and shallow, interspersed by short glides, usually in small groups in linear formation. When feeding keeps in close groups, unlike egrets and herons, sweeping head sideways, with bill submerged. Often sleeps by day, with head on back and bill hidden, but ivory-coloured plumage and closeness of birds to each other aids identification.

Eurasian Bittern *Botaurus stellaris* pm, wv

L 75. W 130. Scarce passage migrant and winter visitor. Occurs mid Oct–May, mostly late Mar–early May. **Habitat** Reedbeds of any size, occasionally damp overgrown ditches and lush riverine vegetation. Occasionally seen in open. **ID** Cryptically patterned brown heron. Smaller than Grey Heron with *stocky neck and entirely dark brown and golden-buff streaked plumage.* In owl-like flight (but with quick wingbeats) neck often extended, before being retracted. Juvenile Black-crowned Night Heron is smaller and shows white spots on wings; juvenile Purple Heron lacks streaks on back and has a long, thin neck. **Voice** Flight call a hoarse *kaau.*

ad
ad
juv
juv
Greater Flamingo
Lesser Flamingo
ad summer
Glossy Ibis
imm
imm
ad summer
ad
Eurasian Spoonbill
Eurasian Bittern

PLATE 11: STORKS AND CRANES

Black Stork *Ciconia nigra* pm

L 95. W 150. Scarce passage migrant. Occurs from Sep, occasionally late Aug–Dec, sometimes to Jan. Fewer Mar–May, mainly Apr. Singles or small groups of up to six are usual. **Habitat** Dams, marshes, brackish pools and fields, mostly seen in flight. **ID** Glossy-black stork with white lower underparts. ***All-black upperparts*** and ***small white axillary patch on black underwing***. Adult has red bill and legs, whereas browner, less glossy juvenile has greyish-green bill and legs.

White Stork *Ciconia ciconia* pm

L 100. W 170. Uncommon and irregular passage migrant. Occurs late Jul–early Nov, mainly mid Aug–early Sep, with stragglers to mid Dec. Occasionally lost juveniles or sick birds in winter. Rare in spring, Mar–May. Usually singles or small groups; major 'falls' of hundreds, even thousands, have occurred mid–late Aug, usually weather-related. **Habitat** Wetlands, plains, irrigated fields and farmland. **ID** Easily told by ***large size, white plumage with black flight feathers, straight red bill and long red legs***. In flight, the neck is extended and legs protrude beyond tail. Juvenile has duller white plumage and duller red bill and legs. **Voice** Clatters bill at nest; otherwise silent.

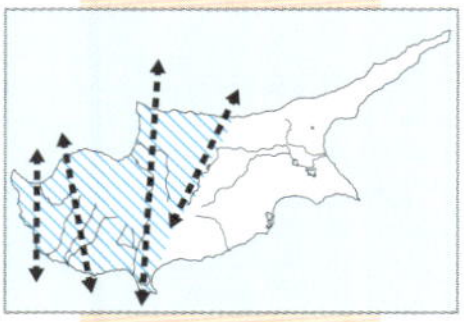

Demoiselle Crane *Grus virgo* PM

L 95. W 175. Fairly common passage migrant. Variable spring passage Mar–mid Apr, mostly mid Mar. In autumn, regular mid Aug–mid Sep, peaking late Aug. Groups rest overnight on Akrotiri Salt Lake. Lost or sick individuals remain occasionally at other times. **Habitat** Salt lakes and dry mudflats, marshes and other shallow wetlands; often seen only in flight, or heard passing over at night. **ID** Smaller with shorter neck and bill than Common Crane, but size deceptive without comparison. Pale grey with largely black head and neck, and elongated breast feathers hanging down in a narrow black fringe. Juvenile grey on head and neck with short whitish band behind eye. Immature gradually acquires adult plumage, general coloration browner with duller black parts. **Voice** Call higher pitched than Common Crane.

Common Crane *Grus grus* PM, wv

L 115. W 233. Common passage migrant and irregular winter visitor. Occurs Feb–early May, occasionally later, mostly Mar–early Apr. In autumn, mainly Sep–Nov (peak 6–8 weeks later than Demoiselle Crane). Variable numbers annually. Often in hundreds (formerly in thousands) and declining. Flocks pass quickly, often not landing. Small groups or singles sometimes present in winter. **Habitat** Salt lakes, dams, cereal fields and farmland. **ID** Large and majestic. ***Grey with contrasting black flight feathers, black head and upper neck, and white stripe from eye down side of neck. Looks 'bushy'*** at rear end on ground. Juvenile has brownish head without contrasting pattern. Adult larger than Demoiselle and ***lacks black breast***. Neck extended in flight, as in other cranes; powerful wingbeats interspersed by long glides, often soars; migrates in flocks, usually in 'V' formation. **Voice** Distant and nocturnal flocks often detected by far-carrying, trumpeting *krrllaa, krrllaa*.

ad
Black Stork
imm
ad
imm
White Stork
ad
ad
juv
Demoiselle Crane
juv
ad
Common Crane

PLATE 12: LITTLE BITTERN AND SMALLER HERONS

Little Bittern *Ixobrychus minutus* PM, ob

L 35. W 55. Common passage migrant and occasional breeder. Occurs mid Mar–early Jun, with a few in summer. Scarce Aug–early Nov, rarely to early Dec. **Habitat** Freshwater marshes, well-vegetated rivers, ponds and dam fringes. **ID** *Small; conspicuous pale covert panels contrasting with black flight feathers and dark back are diagnostic*. Female resembles male but duller with a more rufous tinge, buff streaks on upperparts and brownish streaks below. Juvenile more boldly streaked buff and brown with duller, streaked covert panels. Legs greenish or yellowish. Most often seen in flight, when flushed, usually flying a short distance on rather jerky wingbeats before diving into cover. **Voice** In flight, a short *kek* or repeated *kek-kek-kek-kek*; often calls at dusk. Male in breeding season has loud croaking *khok* repeated at intervals of about two seconds.

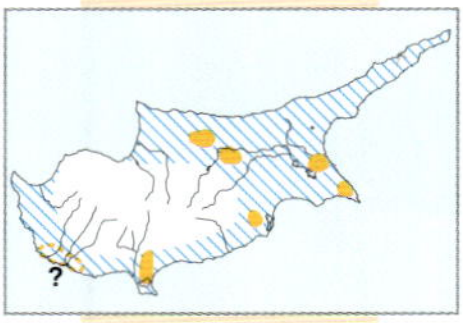

Black-crowned Night Heron *Nycticorax nycticorax* PM, wv, mb

L 60. W 110. Common passage migrant, scarce winter visitor and breeder. Spring Feb–early Jun, mainly late Mar–mid Apr; autumn Jul–Oct, mainly late Aug–Sep. Some in winter. Breeds Apr–Aug. **Habitat** Marshes and ponds with trees and dense vegetation; on passage mainly coastal; nests in large bushes and trees Apr–Aug. **ID** Stocky, about half size of Grey Heron; most active at dawn and dusk. *Adult's grey plumage with black back and crown is unmistakable. Brownish juvenile is prominently spotted with white on back and coverts*, which immediately identifies it from other brown herons. By second calendar year, spotting reduced in size and mantle and scapulars become grey-brown (mirroring black of adult). More likely to be flushed than seen in open during day. **Voice** Harsh deep *kwark*, heard in flight, especially at dusk.

Striated Heron *Butorides striata* v

L 43. W 60. Vagrant. Race *brevipes* has occurred, juveniles most likely. **Habitat** Coasts and harbours, sometimes inland. **ID** *Small, rather dark heron*. Adult has black crown, *bluish-grey upperparts, buff-fringed coverts, and greyish neck and underparts*; yellow patch on lores and dark moustachial streak afford marked facial pattern. Juvenile *brownish with white spots on tips of wing-coverts*, brown- and white-streaked upper breast and yellowish-green legs. May recall young Black-crowned Night Heron, but smaller with dark crown and lacks white spots on mantle. In flight (low with fast wingbeats) all ages show dark upperwing. Often skulking, crouching if disturbed; most active at dusk. **Voice** Alarm call *chook-chook-chook*; when flushed a croaking *kweuw*.

Squacco Heron *Ardeola ralloides* PM, mb

L 45. W 85. Common passage migrant and scarce breeder. Occurs Feb–Nov. Spring passage Mar–mid Jun, occasionally later; in autumn mainly coastal, Aug–Sep, some to Nov. Breeds mid Apr–Jul. **Habitat** Marshy reedbeds, rivers and streams with lush vegetation, and freshwater pools. **ID** Small heron, which usually remains in or near lush cover. Adult in breeding plumage is golden-buff with purple sheen on mantle and *long, black- and white-streaked nape plumes*; bill has greenish-blue base in summer; nape plumes lost in winter, when neck streaked and bill has yellowish base. Juvenile brownish-buff with streaked neck and upper breast, making it well camouflaged. *At all ages, white wings revealed in flight, making bird look predominantly white*. **Voice** In flight, a harsh *kar*.

Western Reef Heron *Egretta gularis* No records

L 60. W 90. Potential vagrant. **Habitat** Beaches, saline lagoons and salt lakes. Feeds from rocks or by slowly wading in shallows at tide's edge, with sudden dash for prey, wings flailing. **ID** Race *schistacea* white morph similar to Little Egret but less elegant, *thicker billed with curved culmen giving slightly drooping look*. Bill is pale brown to yellowish in non-breeder. Legs dark olive-brown, but tarsi often greyer-greenish, with yellowish feet. Non-breeder has facial skin yellow or greenish-yellow. Dark morph *slate-grey with white chin and throat*, occasionally some white flight feathers; juvenile dark morph paler than adult with whitish on foreneck, breast and belly, and can show white in wing, while white morph often has some grey feathers.

♂
♂
♀
ad
juv
juv
Little Bittern
ad
Black-crowned Night Heron
juv
Striated Heron
ad summer
ad summer
ad winter
juv
ad
Squacco Heron
juv
dark morph
Western Reef Heron

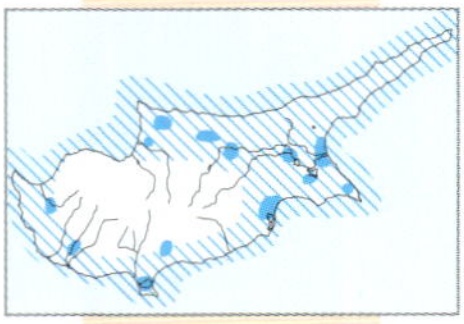

Great Egret *Ardea alba* pm, wv

L 95. W 155. Uncommon passage migrant and winter visitor. Spring late Feb–mid Jun. Most in autumn, mid Aug–mid Nov. Rare in summer and local in winter. **Habitat** Salt lakes, marshes and dams; migration coastal. **ID** ***Largest white heron, with long, angular neck often stretched to full extent***. Non-breeding adult and juvenile have yellow bill and blackish-green or brownish legs. In breeding plumage note scapular plumes, black bill with yellow base and black legs, yellowish above joint. Sedate in flight and on ground. [Alt: Great White Egret, Western Great Egret.]

Grey Heron *Ardea cinerea* PM, WV, ob?

L 95. W 185. Common passage migrant and winter visitor. Occurs all months. Spring passage Mar–mid Jun, peaking mid Mar–mid Apr. Most in autumn, mid Aug–mid Nov, peaking Sep–Oct, when groups coast anticlockwise around island. Probably last bred 1980. **Habitat** Salt lakes, dams and wetlands; migration coastal. **ID** Large with long neck and legs, and powerful bill. In all plumages ***grey upperparts, white underparts and neck***. Adult has ***black crest and dark markings on foreneck***. Juvenile has darker grey or browner upperparts and crown, rest of plumage white. Bill yellowish and legs brownish-yellow in adult, but in younger plumages bill browner and legs greyish. Juvenile can superficially resemble Purple Heron, but is much paler with thicker neck, stouter bill and never shows buffish-brown on neck and breast. **Voice** A raucous *waak* in flight.

Purple Heron *Ardea purpurea* PM

L 80. W 135. Common passage migrant. Occurs mid Mar–mid Jun, peaking early–mid Apr. In autumn mid Aug–Oct, peaking early Sep, when large groups coast south around the island. Scarce in summer and winter. **Habitat** Marshes, reedbeds and ditches. **ID** Slimmer and much darker than Grey Heron, with ***angular, snake-like neck***. Adult dark grey with black belly, ***chestnut on neck edged by black streak, black crown and line on cheeks***; dull yellow bill. Juvenile has sandy-brown upperparts and hindneck, with diffuse dark streaking on neck and blackish crown. ***In flight, has more angular neck (bulging downwards more obviously) than heavier-looking Grey Heron, with prominent spread toes and brownish upperwing-coverts***. Mostly seen in or when flushed from vegetation, unlike Grey Heron which feeds in more open areas.

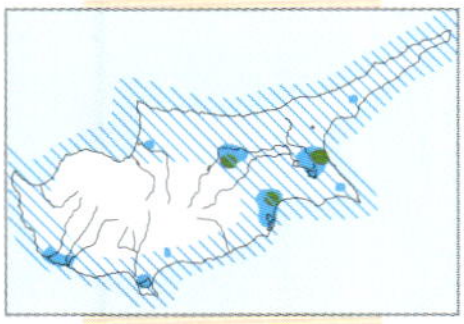

Western Cattle Egret *Bubulcus ibis* RB, PM, WV

L 50. W 85. Common resident, passage migrant and winter visitor. Passage Mar–May, peaking mid Apr; irregular in autumn, mainly late Aug–Oct. Increasingly common breeder, mainly in east. Nests Feb–Jul. **Habitat** Fields and verges, wasteland, usually with grazing livestock; roosts at dams and marshes; nests colonially around wetlands, on bushes or trees. **ID** Small, white, grassland-dwelling heron. Separated from Little Egret by stockier build, ***shorter yellow bill, shorter neck and legs, extended 'jowl' under bill*** and faster flight. Also, in ***breeding season, orange-buff wash on crown, back and breast, and reddish bill***. Usually seen feeding in small groups associated with grazing animals, or flying in flocks to and from roosts. **Voice** A short *ark* and duck-like *og-ag-ag*, often in flight.

Little Egret *Egretta garzetta* PM, wv, mb

L 60. W 90. Common passage migrant, uncommon winter visitor and breeding visitor. Some perhaps resident. Nests May–Aug. Very common on passage Mar–early Jun, peaking mid Apr, and mid Aug–early Nov, peaking mid Sep. **Habitat** Wetlands, often coastal; nests colonially in trees and reedbeds. **ID** Graceful, clean white egret which, in adult plumage, has ***all-black bill and black legs with yellow feet***. In breeding season shows long delicate plumes on nape and mantle. Juvenile has brownish-green legs and pinkish base to lower mandible. From Western Cattle Egret by larger size, bill colour and lack of buffish wash to plumage. **Voice** Throaty grunting *raaak* in flight.

ad winter
ad summer
Great Egret
ad
juv
Grey Heron
ad
Purple Heron
juv
ad winter
Western Cattle Egret
ad summer
ad
Little Egret

Red-throated Diver *Gavia stellata* No records

L 61. W 111. Rare winter visitor to east Mediterranean, and potentially to Cyprus. **Habitat** Mainly coastal. **ID** Long, slender body and bill recalling Great Crested Grebe, but has shorter thicker neck and *upturned head and bill.* In non-breeding plumage, easily confused with Black-throated Diver, but pale grey upperparts speckled white and *pale grey crown and hindneck contrast much less with clean white face and foreneck*; dark red eye conspicuous in white face. *Lacks white patch on rear flanks* of Black-throated Diver, but shows whitish flanks line. Juvenile is duller and darker grey on upperparts, head and neck until moult in late winter. [Alt: Red-throated Loon.]

Black-throated Diver *Gavia arctica* V

L 66. W 120. Vagrant. **Habitat** Coastal waters and large lakes. **ID** Non-breeder more contrastingly black and white than Red-throated, *especially head and neck, with sharp demarcation below eye and on side of neck*. No visible eye-ring makes eye invisible at distance, unlike Red-throated. Head and nape often greyer than nearly black upperparts. *Often shows white patch on rear flanks*. Straight bill held horizontally, *often conspicuously two-toned with blackish tip*. Can show steep forehead. Juvenile has narrow pale edgings to upperparts, but looks blackish at distance. In flight, *best separated by head contrast and black-and-white appearance*, as described for bird on water. May also be confused with juvenile European Shag on sea, from distance.

Northern Gannet *Morus bassanus* pm

L 92. W 170. Scarce passage migrant. Mostly Dec–mid Apr, peaking mid Apr, occasional in autumn. Most are immatures. **Habitat** Maritime. **ID** Large seabird with long pointed bill, cigar-shaped body, and long narrow wings. Adult by *white body and tail, black wingtips and yellowish wash to head*. Juvenile by white crescent on uppertail-coverts and pale speckling, especially above (at close range). With age, white feathers start to appear, producing a patchy black-and-white plumage with *yellow wash on white head from early age*.

Pygmy Cormorant *Microcarbo pygmaeus* V

L 48. Vagrant. **Habitat** Dams, large ponds and marshes. **ID** Small with short neck, rounded head, stubby bill and long tail (useful feature in flight). Bronze-brown head and neck, and glossy black body and wings; in winter, throat is white. Juvenile whitish below with brownish wash on breast.

Great Cormorant *Phalacrocorax carbo* PM, WV

L 90. W 140. Common passage migrant and winter visitor. Small numbers from late Sep or earlier, mainly Nov–Feb, some until Apr. Coastal spring passage mid Feb–early Apr, peaking early Mar. **Habitat** Dams, ponds and reservoirs; coastal especially on migration. **ID** *P. c. sinensis*. Large; swims low in water and frequently perches with wings outstretched. Breeders have *white on nape and neck, and white thigh patch. In winter retains white patch on chin and throat* unlike European Shag. Juvenile browner than adult with dirty white underparts; *from juvenile European Shag by larger size, thicker neck, larger deeper bill and lack of white fringes to wing-coverts*.

European Shag *Phalacrocorax aristotelis* RB

L 75. W 100. Breeding resident. Present year-round, particularly common mid Mar–Apr, especially near colonies. Breeds Nov–May on sea cliffs, rocky islands and ledges. **Habitat** Maritime; perches regularly on cliffs and rocky islands. **ID** *P. a. desmarestii*. Smaller and slimmer than Great Cormorant with thinner neck and bill. Head shape more rounded with *steeper forehead* (in Great Cormorant, rather flat crown merges into bill). All black with green sheen (blue sheen in Great Cormorant), *forehead tuft in breeding season and pronounced yellow gape flanges throughout year*. Plumage duller in winter. Juvenile similar to juvenile Great Cormorant but differs in structure, browner upperparts, whitish-fringed upperwing-coverts, which contrast with dark flight feathers, and whiter underparts with *dark thigh patch*.

ad winter
juv
Red-throated Diver
ad winter
juv
Black-throated Diver
ad
Northern Gannet
imm
ad winter
ad summer
Pygmy Cormorant
ad
imm
imm
ad summer
Great Cormorant
ad
European Shag

PLATE 15: OSPREY AND HONEY BUZZARDS

Western Osprey *Pandion haliaetus* PM

L 56–61. W 145–165. Fairly common autumn migrant, scarce in spring. Occurs late Aug–early Nov, mainly Sep–mid Oct. Scarce late Mar–early May. Sometimes at inland dams in summer. **Habitat** Migration coastal and overland; rests and feeds at dams, reservoirs and salt lakes. **ID** *Large; long narrow wings, distinctly angled when gliding, white undersides with black carpals and band on centre of underwing, white crown and dark mask*. Variable dusky band on foreneck (usually boldest in female). Juvenile has whitish scales and white line on greater coverts above. Flies with steady, shallow wingbeats, glides on smoothly curved wings; may recall large gull at distance. Hovers over water for fish and dives with splash, feet first, almost disappearing.

European Honey Buzzard *Pernis apivorus* PM

L 55. W 135–150. Common passage migrant in autumn, scarce in spring. Occurs mid Aug–Nov, mainly late Aug–Oct, peaking Sep. (Adults migrate late Aug–early Sep, juveniles peak mid Sep.) Light spring passage Apr–mid Jun, peaking late Apr–early May, many passing high over east of island. **Habitat** Widespread on passage, common on Akrotiri Peninsula in autumn. **ID** *Recalls Steppe Buzzard in shape but slimmer with longer, narrower tail with rounded corners; head and neck narrower, protruding in cuckoo-like manner. Wingbeats more flexible; soars on flattish wings and glides on slightly lowered wings* (Steppe Buzzard soars on raised wings). Plumage varies; typical male has greyish head and upperparts; female browner. Below, some dark, others largely white, but *most barred on body and coverts, and have black carpal patches; flight feathers show prominent black trailing edge and characteristic bars at base (more in female). In all morphs, tail has dark band at tip and two at base*. Cere grey, eyes yellow or orange-yellow (male). Juvenile dark brown, rufous-brown or creamy-white with streaked breast; *usually, but not always, with dark carpal patches and narrow whitish crescent on uppertail-coverts; head often whitish with dark mask*; may show pale band on underwing-coverts, separating secondaries from dark forewing, and *three evenly spaced bars on flight feathers* (unlike adult). Juvenile with its more slender wings and curved rear edge (bulging secondaries) and shorter tail has more of a Steppe Buzzard-like outline, but shape of tail and head, and soaring on flat wings important for identification. Migrates in flocks. *See similar Crested Honey Buzzard for separation* from that species.

Crested Honey Buzzard *Pernis ptilorhynchus* V

L 65. W 160. Vagrant. **ID** Resembles European Honey Buzzard but larger with noticeably broader body, slightly longer wingspan, broader wings with bulging secondaries (all ages) and wingtip showing *six long 'fingers'* (five in European Honey), and broader and shorter tail. Dark, pale and intermediate morphs occur (as European Honey). Adult has *dark gorget on throat and lacks contrasting black carpal patches; nape plumes sometimes visible when perched*. Male has *dark red eyes, undertail white, apart from broad black band at tail base, two black bands divided by broad pale band*, and *black band on flight feathers reaches body* (male European Honey has inner part of band hidden by greater coverts). Adult female undertail pattern more like male European Honey, but innermost bar (at tip of tail-coverts) usually broader; secondaries crossed by three dark evenly spaced bars (female European Honey usually has two bars with wider gap between dark trailing edge and first bar). Juvenile has underparts creamy to foxy or dark brown, dark eyes, yellow cere, and variable tail barring, typically four narrow bars of even width, similar to juvenile European Honey, thus best told by shape and structure, but note often has *broad pale rump*, while none shows dark carpal patches (usually obvious in European Honey, except dark individuals). Adult Crested Honey in autumn usually has 4–5 new inner primaries (adult European Honey has 0–3). Care is required with any wintering honey buzzard, as moulting Crested Honey may be missing outer primaries and then only shows five fingers (as European Honey).

ad
ad
juv
Western Osprey
typical
♂
♂
dark morph
♀
juv
typical
♂
♂
pale morph
♂
juv
European Honey Buzzard
♀
♂
juv
Crested Honey Buzzard

Black-winged Kite *Elanus caeruleus* v

L 33. W 76. Vagrant. **Habitat** Open country with perches. **ID** Eastern race *vociferus* has occurred. Small, similar in size to Common Kestrel, but appears larger in flight; also hovers. Plumage essentially ***pale grey and white, with black primaries and large black patch on forewing, contrasting with otherwise pale grey upperparts***. Underparts white; eyes red. Juvenile darker above than adult with white-tipped greater coverts forming narrow line. Note well-protruding broad head, ***shortish tail*** with slight notch when closed, long pointed but relatively broad-based wings. Rather owl-like flight with soft wingbeats; wings raised, rather like a harrier when soaring; persistently hovers. **Voice** A thin whistling *wee-oo, wee-oo*.

Lammergeier *Gypaetus barbatus* v

L 102–115. W 250–280. Vagrant. **Habitat** Bare mountains; more likely to be seen in flight. **ID** ***Outline unique, with long, narrow wings and long, wedge-shaped tail, as well as large size, readily appreciable. Adult has orange-buff underparts contrasting with dark underwings*** and tail; upperparts metallic slate-blue (look black at distance); at close range, golden head has black mask and loose black 'moustache' drooping below bill. Juvenile has blackish-brown head and neck, contrasting with dirty brownish-grey underparts and underwing; largely dark brown above but mantle whitish. Soars and glides effortlessly along mountain ridges on slightly lowered wings; solitary or in pairs. [Alt: Bearded Vulture.]

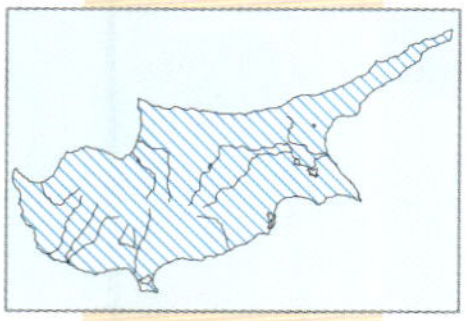

Egyptian Vulture *Neophron percnopterus* pm

L 62. W 155. Scarce and irregular passage migrant. Mostly juveniles and subadults occur. Occurs Sep–mid Nov, peaking late Sep. Rare in spring; historically occurred all months Jan–Jul. Mainly in south, especially Akrotiri Peninsula. **Habitat** On passage, usually only seen in flight. **ID** A small vulture. Adult has ***white wedge-shaped tail***, white underparts with black flight feathers (secondaries greyish-white above), ***small pointed head and thin bill***; colour pattern resembles pale-morph Booted Eagle or White Stork below, but shape quite different. Juvenile is mid-brown below with blackish ruff; dark brown above with creamy bars on wing-coverts, pale rump and whitish uppertail-coverts; ***wedge-shaped tail grey-brown, tipped paler***. Soars on flat to slightly arched wings; in active flight, many deep wingbeats between glides.

Black-winged Kite
ad
juv
ad
Lammergeier
ad
juv
ad
juv
Egyptian Vulture
ad

PLATE 17: VULTURES, SNAKE EAGLE AND BATELEUR

Griffon Vulture *Gyps fulvus* rb, pm?

L 95–105. W 245–270. Scarce and local resident, and possible migrant. Single-brooded; nests late Dec–Aug, fledging from Apr. Reintroduced from Crete since 2012 to restock declining native population. **Habitat** Occurs over all types of country, mainly foothills and mountains in south and west; nests in caves or on cliff ledges. **ID** Large; heavy with long, broad, deeply fingered *wings with curved trailing edge*; short, broad, square-cut tail and slightly protruding narrow head. *Soars effortlessly for long periods on raised wings*; active flight with very slow deep wingbeats; glides on kinked wings. Adult *gingery-buff above and below, contrasting with dark flight feathers*. Juvenile even paler brownish-yellow on rear underwing-coverts, affording even greater contrast with flight feathers. Gregarious. [Alt: Eurasian Griffon Vulture.]

Cinereous Vulture *Aegypius monachus* v, FB

L 105. W 255–295. Vagrant. Extinct as breeder since 1980s. **Habitat** Much as Griffon Vulture. **ID** Very large. From Griffon by all-blackish plumage without any contrast and parallel-edged wings held flat, or slightly downcurved, particularly when gliding; tail also slightly longer and less square-cut than in Griffon. Young blacker than adults, but in both pale legs obvious against black undertail-coverts. At close range adult has black-and-whitish head pattern; head blackish-brown in juvenile. The occasional wingbeat is slow and deep. [Alt: Eurasian Black Vulture.]

Short-toed Snake Eagle *Circaetus gallicus* pm

L 64–73. W 165–180. Scarce passage migrant. Occurs Mar–mid May, peaking early Apr, and Sep–Oct. Scarce other months. **Habitat** Plains, mountains, foothills, but anywhere on migration. **ID** *Large, long-winged eagle with broad head, very pale underparts and square-cut tail with evenly spaced dark bands*. Whitish underparts variably spotted and barred; *some nearly all whitish, others with contrasting dark head and upper breast*; lacks dark carpal patches. Slow, flexible wingbeats, soars on flat or slightly raised wings and hovers regularly. Separated from Western Osprey by broader wings, lack of dark carpal patch and different flight action. Pale-morph Steppe Buzzards and Western Honey Buzzards usually have dark carpals, blacker wingtips, different spacing to tail-bands, and are much smaller with quicker wingbeats. **Voice** Whistling, disyllabic *kee-yo* with long ascending start and short descending finish. [Alt: Short-toed Eagle.]

Bateleur *Terathopius ecaudatus* v

L 55. W 175. Vagrant. **ID** Unmistakable silhouette in effortless flight. Adult in flight has *very short red tail with protruding legs, obvious 'S' curved trailing edge to wings, black head and underparts, chestnut-red back and white underwing-coverts*. Male has black secondaries; female white secondaries, tipped black. Juvenile dark grey-brown with paler head; flight feathers below paler than coverts, dark brown tail longer than in adult and feet do not project beyond tail. Subadult mainly sooty-black but flight feathers a mix of dark and white feathers (all dark in male). Glides fast, 'tilting' on raised wings. When leaving perch, wingbeats fast and light.

ad
ad
Griffon Vulture
ad
Cinereous Vulture
juv
ad
ad dark
morph
ad typical
ad pale
morph
Short-toed Snake Eagle
ad
juv
Bateleur
♂
♀

Lesser Spotted Eagle *Clanga pomarina* pm

L 62. W 145–165. Uncommon and irregular migrant. Occurs mid Sep–early Dec, peaking late Sep–early Oct, usually lone immatures. Rare in spring late Mar–early May, most occurring high over east of island. **Habitat** Anywhere on passage, often near wetlands; in autumn mostly Akrotiri Peninsula. **ID** *Mid-brown adult shows contrast between pale upperwing-coverts and darker brown mantle, underwing-coverts paler than flight feathers, a neat pale patch at base of primaries above* and small creamy area on uppertail-coverts. Juvenile is more likely to occur, and is darker, warm brown below with *flight feathers same shade or slightly darker (coverts never darker than flight feathers)*; unlike adult, may show little contrast between mantle and wing-coverts; darker head has *rusty-yellow spot on nape* (absent in Greater Spotted); narrow white bar on greater upperwing-coverts and distinct whitish inner primary patch. Wings relatively narrow, tail medium-long; active flight less heavy than larger *Aquila* and Greater Spotted Eagles. Soars and glides on arched wings with primaries more or less lowered. On ground lacks heavy 'trousers'. *Both spotted eagles have characteristic round nostrils.*

Greater Spotted Eagle *Clanga clanga* v

L 65. W 155–180. Vagrant. **ID** *Typically darker than Lesser Spotted. Adult shows characteristic single white carpal crescent on underwing (two long, thin crescents in Lesser Spotted). Flight feathers below a similar shade or darker than coverts, resembling older Lesser Spotted; leading underwing-coverts sometimes blackish-brown* (never so in Lesser). On upperwing, mid- to dark brown coverts sometimes contrast with darker mantle (like typical Lesser); no conspicuous pale primary patch above (unlike Lesser). Lacks pale nape spot of young Lesser. *Juvenile is more likely to occur and is blackish-brown below with paler flight feathers* (in Lesser Spotted, coverts brown but flight feathers never paler); *blackish-brown upperwing has one to three white covert bars, often creating pale panel*; large, diffuse primary patch formed by whitish shafts and pale inner primaries (patch smaller, more conspicuous in Lesser). Infrequently, young Greater Spotted is abnormally coloured on body and wing-coverts: (i) '*fulvescens*' type – illustrated; (ii) yellow-brown above and below; (iii) yellow-brown above, normal below; (iv) yellow-brown below, normal above; (v) underwing-coverts greyish, or dark, mottled paler, underparts darker; or normal upper- and underwing, but contrasting paler underparts. *Immatures' secondaries below are finely barred, becoming broader with age* (similar to immature/subadult Lesser Spotted) but disappear in full adult when flight feathers become uniformly dark. Adult has relatively broad and parallel-sided wings with slightly broader 'hand' and deeper 'fingers' than Lesser. Juvenile has narrower 'hand' than adult with trailing edge fairly strongly incurved at body (visible when tail closed). Wing position in flight as Lesser Spotted.

Eastern Imperial Eagle *Aquila heliaca* v, FB

L 72–83. W 190–210. Vagrant. Extinct as resident breeder since at least 1992. **ID** Adult from Golden Eagle by *blackish-brown plumage, contrasting yellow-white hindneck, pale uppertail with broad black band and white 'braces'* (can be hard to see); also, in flight *parallel-edged wings held flattish and, often, closed narrow tail when soaring*. Juvenile has *dark-streaked breast forming pectoral band contrasting with unmarked yellow-buff rear body, and distinct pale wedge on inner primaries below*, yellow-brown upperparts show one or two complete whitish bars on coverts and *creamy lower back and rump. Immature below still streaked* (much as juvenile) *or mottled blackish-brown and yellowish* with rear body clearly paler, *possibly also retains pale inner primaries*; adult head and tail pattern appear early. Rather long-winged with ample 'hand', deep-fingered wingtip (long seventh primary), well-protruding head and relatively long tail. Juvenile, which is more likely to occur, has broader, more 'S'-curved rear edge to wings. Wings sometimes slightly lifted when soaring, but arched during fast glides. Perched juvenile/immature shows *pale lower underparts* and, like adult, rather long protruding head (compared to other *Aquila*).

ad
juv
ad
Lesser Spotted Eagle
juv
ad
ad
'fulvescens'
Greater Spotted Eagle
juv
ad
sub ad
Eastern Imperial Eagle
ad

PLATE 19: LARGE BROWN AND SMALLER EAGLES

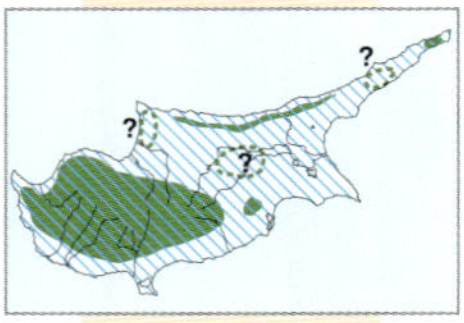

Bonelli's Eagle *Aquila fasciata* rb, pm?

L 60–70. W 150–165. Uncommon resident. Most likely to be seen late Mar–May and Sep–mid Jan. Breeds Jan–Jun; flying young from May. Reports of passage lack evidence. **Habitat** Breeds in forests, lightly wooded foothills and ravines; non-breeders wander, often near wetlands. **ID** In flight, recalls large, thickset honey buzzard. *Adult identified by dark underwings contrasting with whitish underparts, pale tail with black band at tip, and white patch on mantle*; at close range *white leading edge to wing*. Pale rusty-buff juvenile lacks black tail-band, and flight feathers are pale with fine dark barring; *paler translucent primaries contrast with blackish wingtip*; if present, *narrow dark bar on rear underwing-coverts diagnostic*, but can be confined to dark 'comma' on primary-coverts; upperwing cinnamon-brown with large, pale primary patch. Soars on flat or slightly arched wings, often with long, almost square-cut tail held closed (may be twisted independently); glides with carpals pressed forward, trailing edge of wings straight (recalling European Honey Buzzard). Often hunts in pairs; stoops at great speed.

Booted Eagle *Hieraaetus pennatus* pm

L 43–53. W 110–130. Uncommon autumn passage migrant, rare in spring. Occurs late Aug–early Nov, peaking late Sep–early Oct. In spring mid Mar–Apr, peaking probably late Mar, when most pass high over east of island. Rare in winter and summer. **Habitat** Anywhere on migration, usually open country. **ID** Two distinct colour morphs. Size of Steppe Buzzard, but outline and wing position like Black Kite; *tail square-cut*. More ample, deeply 'fingered' wings than Steppe Buzzard. Pale morph has *creamy-white underparts with contrasting blackish flight feathers*, kite-like panel on upperwing, *pale scapulars (seen head-on as 'landing lights') and uppertail-coverts, and diagnostic white spots at base of neck. Lacks dark carpal patch of most pale Steppe Buzzards and European Honey Buzzards*; also has darker base to flight feathers and paler inner primaries. Dark morph similar above to pale morph but underparts dark brown, or rufous with black band on centre of underwing. When perched, *feathered tarsi also separate Booted Eagle from these and Long-legged Buzzards*. Deeper, more powerful wingbeats and steadier glides than Steppe Buzzard; soars on flat wings; does not hover.

Golden Eagle *Aquila chrysaetos* v

L 78. W 190–230. Vagrant. **ID** Powerful flight with *flexible wingbeats; soars and glides on obviously raised wings, with fairly long tail and slightly 'S'-curved rear edge to wings* (more pronounced in juvenile). Dark brown adult has *rusty-yellow hindneck, pale panel on upperwing-coverts*, dark-barred, black-tipped greyish flight feathers which show as *greyish area on outer wing above*, and greyish tail with blackish band at tip. Adult Eastern Imperial Eagle is blacker above, including outer wing, has flatter, more parallel-edged wings when soaring and narrower tail. *Juvenile and immature have white patches in primaries and inner tail, the latter with broad black band at tip*, unique among *Aquila* eagles. Birds older than one year show pale panel on upperwing-coverts.

juv
ad
ad
Bonelli's Eagle
Booted Eagle
ad
pale morph
ad
pale morph
ad
dark mcrph
ad
juv
ad
Golden Eagle
sub ad

Levant Sparrowhawk *Accipiter brevipes* pm

L 32–39. W 65–75. Rare passage migrant. **Habitat** Over open country with trees and bushes; headlands and peninsulas, mainly in south. Most occur mid Sep–early Nov, peaking early Oct. Rare Apr–early May. First-winters predominate in autumn. May occur anywhere, but most seen on Akrotiri Peninsula and at Cape Greco. **ID** Below, ***male has white underwing with contrasting black tip extending to inner primaries***. Female, with more distinctly barred underparts, has less contrasting dark wingtip. ***Black throat-streak, greyish cheeks and lack of supercilium separate from Eurasian Sparrowhawk***; also by ***long pointed wings***, shape of wingtip (four free outermost primaries, five in Eurasian Sparrowhawk), and often by more rounded tail corners. Adult female has dark outermost primaries (absent Eurasian Sparrowhawk) and lacks female Eurasian Sparrowhawk's fully banded central tail feathers above. Juvenile Levant is dark grey-brown above with closed tail banded, though often indistinctly; ***underparts have dark longitudinal spots, almost forming lines on breast***, and dark throat-streak.

Eurasian Sparrowhawk *Accipiter nisus* PM, WV, ob

L 29–40. W 60–80. Common passage migrant, winter visitor and occasional breeder. Autumn passage late Aug–early Dec, peaking mid–late Oct. Lighter spring passage Mar–Apr, continuing some years to May or Jun (when nesting may occur). Variable numbers in winter. **Habitat** Open country with trees at all elevations; occasionally forest edges when nesting. **ID** Female much larger than male, approaching male Northern Goshawk in size, but wingbeats faster and lighter, body slimmer, with less protruding head, wingtips blunter and tail thinner, longer and more square-cut. Adult ash-grey (female) or bluer slate-grey above (male), barred rufous or brown below; whitish supercilium in female (infrequent in male); ***pale underwing without dark tip***. Juvenile browner above with clear white supercilium; streaked or blotched throat and upper breast, otherwise barred below. Quick wingbeats interspersed by short descending glides (stronger, straighter glides in Northern Goshawk); display flight has slow harrier-like wingbeats, also occasionally when hunting.

Northern Goshawk *Accipiter gentilis* rb, pm

L 48–60. W 90–125. Rather elusive resident; a few on passage and possibly in winter. Breeds mostly in Paphos, Troodos and Makheras forests, with some wandering to nearby foothills and plains year-round. Passage Mar–May, peaking early Apr; and Aug–early Nov, peaking early Oct. Breeds Mar–Jul. **Habitat** Prefers coniferous forest, less often open country. Breeds in damp, remote forest valleys. **ID** ***Female much larger than male, with wingspan of Steppe Buzzard. Compared to female Eurasian Sparrowhawk, male has deeper belly, slower stronger wingbeats, and longer, broader-based but more pointed wings***. Note shorter, broader-based tail, with rounded corners (thinner tail more square-cut in Sparrowhawk) and ***more protruding head and neck***. Stronger straighter glides than Sparrowhawk and may soar on upturned wings. Adult dark grey above, darker head appears 'hooded' but supercilium white; white underparts finely barred. Juvenile dark brown above with pale mottling on ear-coverts; rusty-yellow ***underparts boldly streaked darker***, lacks 'hooded' appearance of adult. Female from large falcons by more rounded wings, bold tail-bands, and flight. Hunts like Eurasian Sparrowhawk but also runs down prey on ground; display flight of shallow harrier-like wingbeats. Elusive and inconspicuous in forest habitat, where spends much time perched in trees or hunting among them. Treated warily by crows.

Levant Sparrowhawk

Eurasian Sparrowhawk

Northern Goshawk

Western Marsh Harrier *Circus aeruginosus* PM, WV

L 48–55. W 115–130. Common passage migrant and winter visitor. Autumn passage Aug–early Nov, some overwintering at wetlands. Spring Mar–May; rather scarce in summer. Immatures predominate. **Habitat** Marshes, reedbeds and coastal farmland. **ID** Larger and broader winged than other harriers; wavering low glides on raised wings when hunting. *Male has tricoloured wings*; underwing white but wingtip black and *rear body red-brown*. Female dark brown with *crown, throat and breast spot yellowish-white*. Juvenile blackish-brown, usually with rusty-yellow head. Immature male has dirty grey areas on upperwing and tail, rusty-brown body and underwing-coverts, and more extensive black wingtips than adult. Rare dark morph sooty brown but adult male has distinct white base to flight feathers below.

Hen Harrier *Circus cyaneus* WV, pm

L 45–56. W 100–120. Winter visitor and passage migrant. Occurs Sep–early May, with variable numbers in winter. Immatures predominate. **Habitat** Marshes, farmland and meadows. **ID** Slimmer than Western Marsh Harrier with more buoyant flight. Male has *clear-cut white uppertail-coverts, uniform pale grey upperparts, head and upper breast, and extensive black wingtips*. Second-autumn male can show black wedge on wingtip like male Pallid Harrier (due to primary moult). Female and juvenile brownish with white uppertail-coverts; streaked underparts whitish or rusty-yellow (warmest in juvenile); banding on secondaries below most distinct in female. Juvenile has pale tips to greater upperwing-coverts and is best separated from juvenile Montagu's and Pallid by proportionately *shorter, broader wings with more rounded wingtip* (formed by four outermost primaries – three in the other two species), *less buoyant flight and streaked breast* (unstreaked rusty-yellow in juvenile Pallid and Montagu's).

Pallid Harrier *Circus macrourus* PM

L 40–48. W 95–117. Fairly common passage migrant. Spring Mar–early May; less common Aug–Oct. Occasional in winter. Immatures predominate. **Habitat** Marshes, foothills and farmland. **ID** Proportions and flight similar to Montagu's Harrier. Male pale grey above *without clear-cut white rump; whitish head and underparts with black wedge on wingtip*. Female from Montagu's by *pale, dark-streaked, collar* (like female Hen Harrier), *less space between dark bands on secondaries below, with pale bands becoming darker towards body*; primaries below often pale, contrasting with darker secondaries and without distinct dark trailing edge; heaviest barring on central primaries with *bases often unbarred, creating pale 'boomerang' surrounding darkish coverts*; distal primaries with faint or no barring, except narrow dark 'fingertips' to longest primaries (unlike Montagu's). Except pale leading arm, *most underwing-coverts and axillaries rather dark-streaked* and lack distinct pattern (boldly rufous-barred on Montagu's). Streaks on underparts largely confined to upper breast, which contrasts more with paler rear body than in Montagu's. Juvenile has *broad pale collar bordered by brown neck*; primaries below rather evenly barred from base to tips, though *often has pale 'boomerang' at primary bases, 'fingers' never all dark as in most young Montagu's*. Young male has paler head and breast than Montagu's; new central tail feathers show diffuse dark bands near tips (Montagu's has grey neck and breast, contrasting with paler belly, and new central tail feathers plain grey).

Montagu's Harrier *Circus pygargus* PM

L 43–47. W 97–115. Fairly common passage migrant. Occurs mid Mar–mid May and more commonly mid Aug–Nov. Immatures predominate. **Habitat** Marshes, foothills and farmland. **ID** Slender build, narrow wings and buoyant flight. Male has grey back and inner wing, silvery-grey outer wing with extensive black wingtips; one black band on secondaries above and two below; red-brown streaks below dark grey upper breast. Second-autumn male can show black wedge at wingtip due to primary moult (recalling male Pallid Harrier). Female has rufous-streaked underparts, well-spaced dark bands on pale secondaries and evenly barred primaries from base to tips, with dark trailing edge to 'hand'; close to, pale underwing-coverts and axillaries show uniform bold rufous bars. Juvenile dark rufous to yellowish-ochre below, largely unstreaked; lacks distinct pale collar of young Pallid; 'fingers' and trailing edge of 'hand' below dark, but 'hand' otherwise pale with fine regular barring from base to tips. Rare melanistic morph sooty-black with pale base to primaries below.

♂
♀
♂
Western Marsh Harrier
♂
♀
♂
Hen Harrier
juv
♂
♀
♂
Pallid Harrier
juv
♂
♀
♂
Montagu's Harrier
juv

Red Kite *Milvus milvus* V

L 60–70. W 145–170. Vagrant. **ID** Larger and slimmer than Black Kite with ***longer, deeply forked, rusty-red tail, prominent white patch on primaries below and more buoyant flight***. In adult, underparts largely dark rusty-red, contrasting with pale undertail; broad buffish-brown band on inner wing above. Juvenile has cinnamon breast with bold pale streaks, rest of underparts and undertail pale; duller red-brown uppertail has slight, dark subterminal-band, broader, paler band on inner wing above and ***greater upperwing-coverts show thin white line***. Light and graceful soaring and gliding on arched wings, and constantly manoeuvred tail (looks only slightly forked when fully spread).

Black Kite *Milvus migrans* PM, wv

L 50–65. W 125–150. Fairly common passage migrant. Autumn late Aug–Oct, mainly mid–late Sep. Uncommon in spring mid Mar–mid May, mainly Apr–early May. Spring migration possibly overlooked, when numbers pass high over east of island. Occasional in winter. **Habitat** River valleys, farmland, salt lakes, often near water; anywhere on migration and often at rubbish dumps. **ID** *M. m. migrans* generally occurs, but many autumn and particularly winter birds in Cyprus show intermediate features between *migrans* and *lineatus* ('Eastern Black Kite'). Resembles Red Kite in silhouette but ***tail less forked*** (often square-ended when fully spread), slightly broader wings and ***less buoyant flight***. Adult darker and browner than Red Kite, with ***hardly any white on primaries below and dark brown (not reddish) uppertail***; head darker and pale panel on upperwings-coverts less conspicuous. Juvenile shows dark mask, pale feather tips to mantle and shoulders, boldly dark-spotted breast but paler belly and diffuse dark band to tail; also ***whitish tips to greater upperwing-coverts***. Elastic wingbeats, soars and glides on slightly arched wings; manoeuvres tail when scanning for food. Gregarious at rubbish dumps and at roosts outside breeding season. In *migrans* feet and cere/base of bill are bright yellow, but these parts are chalk-white to pale bluish in *lineatus*, at all ages and plumages. *See Black-eared Kite for other features*. **Voice** Gull-like, a high-pitched, whinnying, descending, drawn-out *yiieerr*.

Black-eared Kite *Milvus* (*migrans*) *lineatus* No records

L 55–68. W 135–162. May occur. **Habitat** As Black Kite. **ID** Very similar to Black Kite, but many are slightly larger. Adult Black-eared differs in having ***white bases to outer primaries, broader 'hand' with long sixth primary, all six 'free' primaries longer than in Black Kite and browner head (greyish in Black Kite) with more obvious dark mask***. Underparts streaked ochre (finer black streaking in Black Kite) and bases to inner primaries noticeably barred dark and white. Juvenile similar to juvenile Black Kite but has broader 'hand' with long sixth primary (like adult), whiter base to primaries, more noticeable streaking on breast (not unlike that of juvenile Eastern Imperial Eagle) and pale belly and vent.

White-tailed Eagle *Haliaeetus albicilla* V

L 80–90. W 190–250. Vagrant. **Habitat** Wetlands, dams, lakes and coasts. **ID** Large and bulky; adult has broad, parallel-edged wings, ***short white wedge-shaped tail, long protruding neck and large yellow bill***. Juvenile has curved trailing edge to wings, broader 'arm' and narrower 'hand' than adult; appears dark at distance, but at close range ***head and neck blackish***, contrasting with rusty-brown, dark-streaked underparts; closed tail looks dark but when spread shows ***whitish feather centres***; axillary patch and lines on central underwing-coverts also whitish and has pale loral patch. Immature mottled whitish below and on mantle and shoulders, but still dark-headed. Soars and glides on flattish or slightly arched wings; in active flight a long series of heavy wingbeats. **Voice** Loud, ringing *klee-klee-klee*.

ad
juv
Red Kite
ad
juv
Black Kite
juv
Black-eared Kite
ad
juv
ad
White-tailed Eagle

PLATE 23: BUZZARDS

Rough-legged Buzzard *Buteo lagopus* V

L 50–61. W 125–145. Vagrant. **ID** Confusable with slightly larger Long-legged Buzzard; note prominent dark tail-band and lack of dark line on underside of greater coverts. Larger and heavier than Steppe Buzzard, wings and tail longer; ***wingbeats flexible and wings kinked when gliding. Separated from all other buzzards by white tail with broad blackish band near tip***. In adult, tail-band narrowest and most clear-cut, with two to four extra bars in male (none or one in female). ***In juvenile, tail-band less clear-cut and inconspicuous from below***. Adult typically pale-headed with dark upperwings; pale 'U' on breast separates dark belly from dark foreneck. ***Large dark carpal patches and tarsi feathered to toes at all ages***. Juvenile has pale band on paler upperwing-coverts, larger pale primary patch above and is whiter below; lacks clear-cut dark trailing edge on underwing. Routinely hovers.

Common Buzzard *Buteo buteo buteo* WV, pm

L 50. W 120. Common winter visitor and uncommon passage migrant. Mainly Oct–mid Apr. Many birds show characteristics of both *B. b. buteo* and *B. b. vulpinus*, some being inseparable. Due to the location of Cyprus, it is thought that the '*intermedius*' form (intergrade between *vulpinus* and *buteo*) may predominate on passage and occur in winter. **Habitat** Usually at low elevation, including farmland, river valleys, light woodland and reedbeds. **ID** The well-known buzzard of Western Europe. Similar to grey-brown morph of Steppe Buzzard, but thicker-set with broader, longer wings and slightly shorter tail.

Steppe Buzzard *Buteo buteo vulpinus* PM

L 48. W 118. Common passage migrant and rare winter visitor. Scarce mid Feb–early May. Mainly early Aug–Nov, peaking late Sep. Passage 10–15 days later than European Honey Buzzard. Rare in winter. **Habitat** Anywhere on passage, most in autumn on Akrotiri Peninsula. **ID** Slightly smaller than Common Buzzard with narrower, more pointed wings and longer tail. Three main morphs occur: fox-red, grey-brown and blackish (scarce). Fox-red and blackish morphs very similar to Long-legged Buzzard, but smaller with ***shorter, narrower wings, shorter tail, stiffer wingbeats and glide on flattish wings*** (Long-legged Buzzard has flexible wingbeats and kinked gliding profile); also Steppe Buzzard's wings are less raised when soaring. Fox-red morph has more variegated rusty-orange underwing-coverts framed by dark greater coverts, with narrow blackish comma-shaped carpal patch (usually large in Long-legged); rufous-brown uppertail more barred (but can be plain like Long-legged), head dark and pale primary patch above usually small. Blackish morph from Long-legged by flight and silhouette. Grey-brown morph resembles Common Buzzard. Juvenile has streaked breast, diffuse band on trailing edge of underwing, no dark band on tip of tail; it is often pale-headed, with pale upperwing-coverts and prominent primary patch above; from Long-legged by structure. Young blackish morph has fine, regular barring on secondaries below and on uppertail (coarser, more irregular in similar Long-legged). Soars on raised wings and may hover when hunting. Migrates in flocks.

Long-legged Buzzard *Buteo rufinus* RB, pm, wv?

L 60–66. W 130–155. Common resident and scarce passage migrant. Has expanded across island since first breeding record in 1992. Breeds late Feb–mid Jun and immatures wander. Some migration reported Mar–early May and late Aug–Oct. **Habitat** Scrub-covered foothills, valleys and light woodland. **ID** Larger than Steppe Buzzard with *longer wings and tail*, kinked wings when gliding and flexible wingbeats; ***soars on raised wings***. Much plumage variation: creamy-white, rufous-brown and blackish morphs based on colour of body and underwing-coverts. Blackish morph can show coarse dark bars on flight and tail feathers. ***Typical Long-legged Buzzard has pale head and breast, becoming dark on belly, pale sandy or rufous-brown upperwing-coverts contrasting with flight feathers, unbarred pale rusty-orange uppertail and large black carpal patches***. Juvenile has finely barred outer tail and diffuse dark trailing edge to underwing; some almost white below with bold carpal patches and dark rusty-brown belly or sides. Perches prominently, soars or hovers when hunting. **Voice** Mewing *peeeoo*, similar to Steppe Buzzard.

Rough-legged Buzzard
♂
♀
juv
blackish morph ad
fox-red morph ad
ad
Common Buzzard
Steppe Buzzard
fox-red morph juv
grey-brown morph imm
ad
juv pale morph
juv dark morph
Long-legged Buzzard

Great Bustard *Otis tarda* V

L 80–100. W 190–260. Vagrant. **ID** Huge, with upright neck, ***tail sometimes cocked. Male develops brown-red breast-band with age*** and has white whiskers in breeding season. Twice the size of female. Female and juvenile have slimmer neck without breast-band. ***Flight powerful with deep wingbeats***; broad wings, outstretched neck and short tail; ***large white wing-patches*** very conspicuous. Shy.

Macqueen's Bustard *Chlamydotis macqueenii* V

L 60. W 150. Vagrant. **Habitat** Stony plains, salt flats and anywhere on passage. Now separated from Houbara Bustard *C. undulata*, which occurs across North Africa. **ID** A large bustard, with ***black frill on side of neck in all plumages***. Flight rather slow, wingbeats shallow, with ***long-tailed and narrow-winged appearance and white patch confined to outer primaries***. Shy, prefers sneaking away without flying.

Little Bustard *Tetrax tetrax* V

L 43. W 110. Vagrant. **Habitat** Grassy meadows, large cereal and fodder fields. **ID** Relatively small, with ***small head and long neck***. Male has ***grey, black and white neck pattern***; non-breeding male, female and juvenile lack striking neck markings. Flight fast with rapid, stiff, winnowing wingbeats showing ***almost all-white wings with black largely confined to outermost primaries***. Rather shy, frequently in cover of grass or low vegetation. In flight, male's wings make whistling noise.

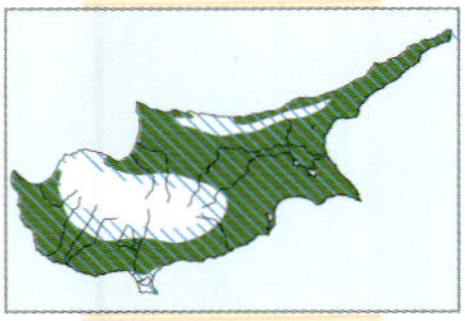

Eurasian Stone-curlew *Burhinus oedicnemus* RB, PM, wv?

L 42. W 81. Common breeding resident, passage migrant and possible winter visitor. Spring passage mid Mar–mid May, peaking mid Apr. Autumn passage Aug–Oct, coinciding with formation of post-breeding roosts. Nests Mar–Jun, calling to Sep. **Habitat** Roosts in loose groups in ploughed fields, olive groves, salt scrub, waste ground and dry stony riverbeds, where it also nests. **ID** *B. o. saharae*. Large, streaked, curlew-coloured wader with ***short bicoloured bill and large, staring yellow eyes***. In dry habitats, often 'freezes' motionless or walks slowly in hunched posture. Flies with ***stiff wingbeats and shows two small white 'windows' in primaries and a paler wing panel, bordered in front by dark and white lines***. Often in flocks. More active by night than day. **Voice** Vocal mainly at night, reminiscent of Eurasian Curlew's *cur-lee* with emphasis on higher pitched second syllable; also loud Oystercatcher-like *ku-beek, ku-beek*.

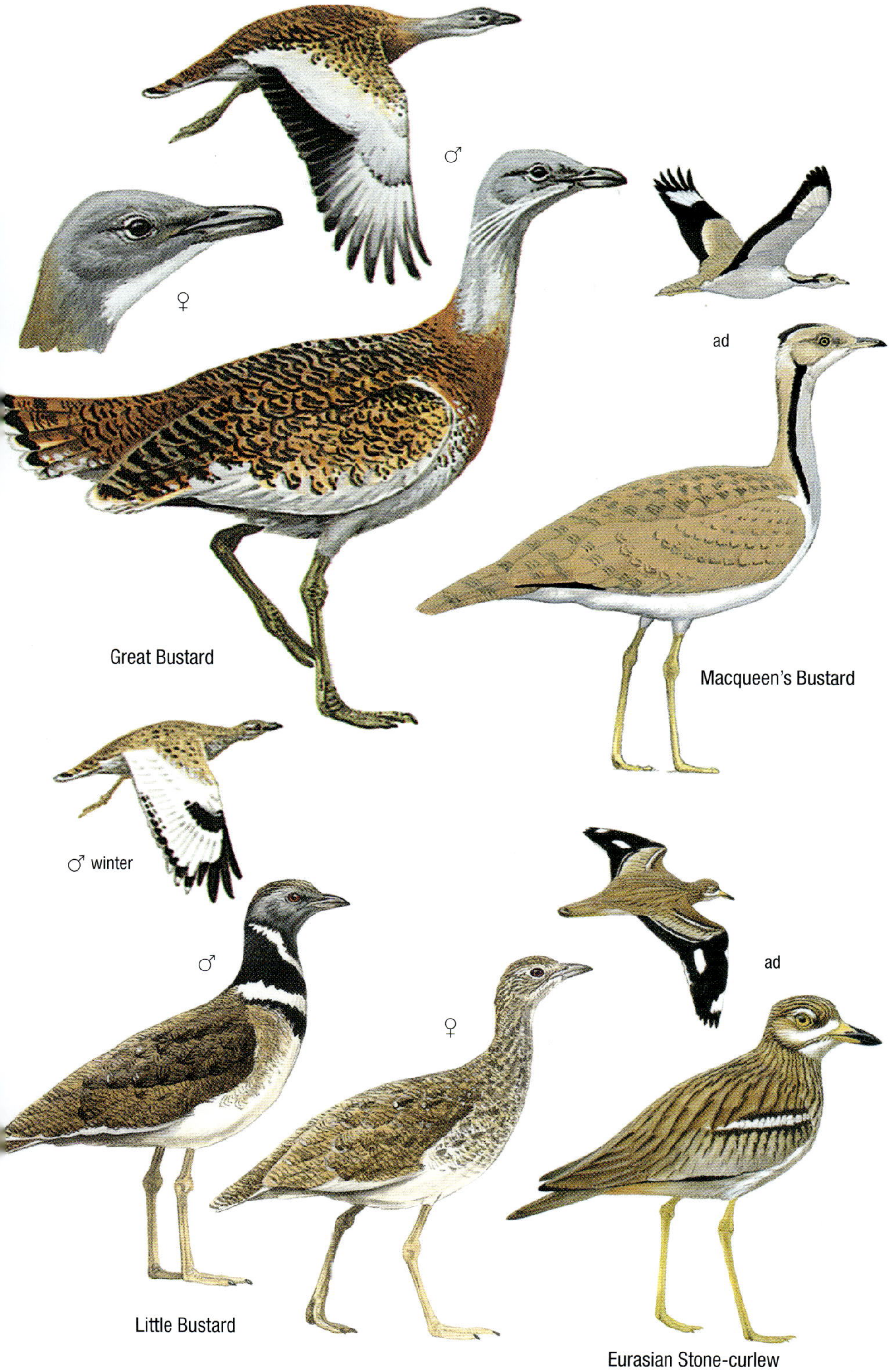
♂
♀
Great Bustard
ad
Macqueen's Bustard
♂ winter
♂
♀
Little Bustard
ad
Eurasian Stone-curlew

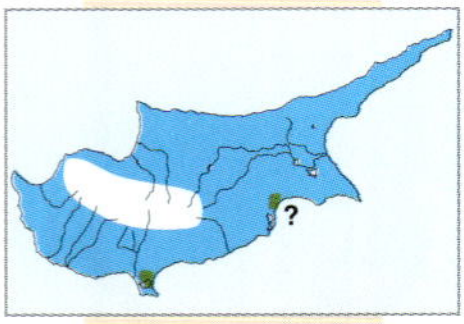

Water Rail *Rallus aquaticus* WV, PM, ob

L 26. W 41. Fairly common winter visitor and passage migrant; occasional breeder. Migrants occur mid Aug–mid May. Passage Mar–mid May and Oct–Nov. Breeds May–Aug. **Habitat** Reedbeds and marshes, waterlogged meadows, sewage pools, wet ditches and riverine vegetation. **ID** Secretive; *larger than* Porzana *crakes* and easily distinguished by *long red bill*. Adult has dark, *mottled brown upperparts*, uniform slate-blue head-sides, *heavy black and white bars on flanks and conspicuous white undertail*. Juvenile has browner face, mottled grey-blue underparts and blackish bill. Tail often cocked and jerked when walking. When flushed, flies only short distance on fluttering wings with long legs dangling. **Voice** Grunting, groaning, whining and *stomach-churning sounds*; sometimes *like a squealing pig*. In spring, utters for hours a rhythmic *trüt-trüt, trüt*, sometimes ending in a trill.

Little Crake *Porzana parva* PM, wv

L 19. W 37. Common passage migrant. Spring Feb–early May or later, peaking Mar–Apr; scarce Aug–Nov, occasional winter. **Habitat** Marshes, reedbeds, slow-moving streams, boggy fields, sewage pools and vegetated ponds. **ID** Smaller than Spotted Crake and separated by *uniform blue-grey underparts (adult male) and heavily barred undertail-coverts*. Less compact than Baillon's with longer legs and neck, and *longer wing projection*. Tertial pattern differs: pale buff fringes to inner webs form *broad creamy line along inner aspect of folded wing* (Baillon's paler edges to tertials never form broad continuous line). Male from Baillon's by *less barred flanks*, less spotted upperparts, red base to bill, and green legs. *Female has brown-buff underparts*, white chin and throat, and some grey on cheeks and supercilium. Juvenile lacks grey head pattern and has stronger flanks barring than adult; from juvenile Baillon's by *structure, tertial pattern* and less barred underparts. Any small crake with buff underparts seen in winter and spring will be female Little Crake. **Voice** Seldom heard. Male's song a loud, accelerating croaking *kwak… kwak… kwak, kwak, kwak-kwak-kva-kva-kva-kva-kva*.

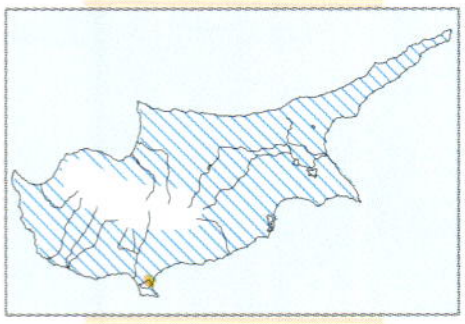

Baillon's Crake *Porzana pusilla* pm, ob

L 18. W 35. Scarce passage migrant and occasional breeder. Mainly late Mar–mid May and Aug–early Nov. Breeds May–Jul. **Habitat** Reedbeds, marshy pools (where also breeds) and slow-moving rivers. **ID** Resembles Little Crake but *more compact with very short primary projection*. Upperparts warmer rufous with distinct, but small, irregularly scattered white spots; *tertials do not form continuous pale line along inner aspect of folded wing* as in Little Crake; underparts bluish-grey in both sexes with *heavier black and white barring on flanks. Uniform green bill without red base*, and dirty olive legs and feet. Juvenile more strongly barred below than Little Crake, best told by short primary projection and absence of broad, pale tertial line. Most skulking of all the crakes. **Voice** Seldom heard. Song a series of dry, rattling frog-like sounds lasting 1–2 seconds and repeated at intervals of 1–2 seconds, *trrrrr, trrrrr, trrrrr*.

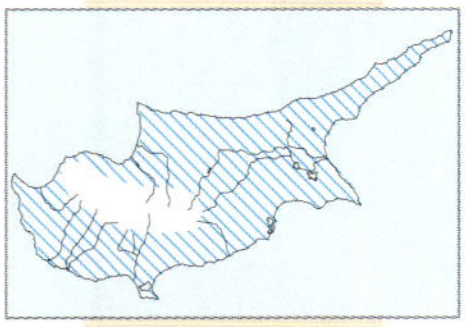

Spotted Crake *Porzana porzana* PM

L 23. W 40. Fairly common passage migrant. Occurs Mar–May, mainly late Mar–Apr; autumn Aug–Nov, mainly mid–late Oct, occasionally winter. **Habitat** Reedbeds, marshes, ditches, pond margins and rivers. **ID** Small, round-bodied crake, larger than Little Crake. *Heavily white-spotted plumage* and short red-based yellow bill; flanks and vent barred black and white, but *buff undertail-coverts* (barred in Little and Baillon's) visible when walking with tail cocked. Juvenile lacks grey head pattern, has whitish throat and brownish underparts with whitish spots. Secretive, moves with slow, stalking steps and sudden crouching run. In flight dangles legs. **Voice** Song a *rhythmical, far-carrying whistle whitt, whitt*, repeated at c.1-second intervals, mainly in night from dusk.

Corn Crake *Crex crex* pm

L 29. W 50. Uncommon passage migrant late Mar–mid May and late Aug–early Nov. **Habitat** Cereal fields and irrigated farmland, but many found dead hit by cars, taken by cats or shot by hunters. **ID** Skulking, rarely leaves cover unless flushed; larger than Common Quail. In flight suggests young gamebird, but note *chestnut wing-coverts* and often dangling legs. Otherwise located by *characteristic song*, occasionally heard on passage. **Voice** Far-carrying, disyllabic rasping, *arrp-arrp, arrp-arrp*.

ad
juv
ad
Water Rail
♂
♀
juv
ad
Little Crake
ad
juv
ad
Baillon's Crake
ad
juv
ad
Spotted Crake
ad
Corn Crake

Grey-headed Swamphen *Porphyrio poliocephalus* V

L 47. W 95. Vagrant. **ID** Large and heavy; twice the size of Common Moorhen. ***Uniformly bluish-purple with grey head and neck, huge red bill and frontal shield; red legs and long toes***. Swims with body stooped forward, ***white undertail-coverts often striking***. In flight, like huge Common Moorhen but with blue wings and long red legs. Juvenile drabber, with greyish-blue underparts, dull red legs and greyish bill. Shy and often in cover. **Voice** Loud clucks, clanks, bugling and deep mooing notes; a low *chock-chock*; also *tschak-tschak* and nasal bugled song *quin quin krrkrr, quin quin krrkrr*. [Alt: Purple Gallinule, Purple Swamphen.]

Allen's Gallinule *Porphyrio alleni* V

L 23. W 50. Vagrant. **ID** Smaller, more elegant than Common Moorhen, with bluish head and underparts, and ***iridescent greenish wings. Lacks white flanks line***, but has white undertail-coverts. ***Note red legs and greenish upperparts***. Juvenile/first-winter has warm brown upperparts, distinctly pale-fringed tertials and greenish wash to flight feathers; underparts buffish with white belly and undertail-coverts; frontal shield brownish; legs brownish turning red. Jerks tail when moving, and walks easily over floating plants; swims well. Long legs dangle in short flight. Secretive.

Common Moorhen *Gallinula chloropus* PM, WV, RB

L 33. W 52. Common passage migrant, winter visitor and resident. Spring passage mid Mar–mid May, peaking early–mid Apr, and autumn passage mid Sep–mid Nov, peaking mid Oct–early Nov. Winter visitors present at least Dec–mid Feb. Irregular and scarce breeder prior to 1992, but now common. **Habitat** Sheltered wetlands including marshes, ponds, reedbeds, perennial streams and dams with dense vegetation and nearby grazing; clambers and hides in waterside bushes. **ID** Dark, hen-like waterbird, with prominent ***red bill and shield, and white flanks line. Constantly jerks tail to show white undertail-coverts***. Swims with vigorous nodding motion and body stooped forward. Juvenile paler, grey-brown with dark bill; from young Eurasian Coot by ***white flanks line and undertail pattern***. Seeks cover and patters over water when disturbed; often walks in open in wet grassland, marshes. **Voice** Sings at night with persistent clucking *kreck-kreck-kreck-*.... Many calls can be confused with those of Eurasian Coot; sometimes a short variable *kek or kr-r-eck*; also characteristic a sudden, loud, gurgling *grrll* and a two- or three-note *kwett – kwette-wett*.

Eurasian Coot *Fulica atra* WV, PM, RB

L 36–42. W 75. Very common winter visitor, common passage migrant and resident. Locally in large flocks. Migrants and winter visitors supplement residents Sep–early May, peaking mid Dec–Jan, and some may stay to breed. Family parties widespread Apr–Aug. **Habitat** Larger freshwater wetlands, including dams, reservoirs, water treatment plants and reedy marshes. **ID** ***Adult sooty-black with white bill and frontal shield***; hunchbacked on water. Upright stance out of water when note ***long greenish legs and lobed feet***. Flight stronger and heavier than other rails, more duck-like, on rounded wings with long pattering run across water before take-off, and ***narrow white trailing edge on inner wing***. Long toes trail beyond tail. Juvenile duller and paler with nearly white underparts and smaller frontal shield. Dives well, but only briefly. Gregarious, especially in winter. **Voice** Commonest call a short staccato *kewk*, also an explosive high *pitts*.

juv
ad
Grey-headed Swamphen
ad
juv
Allen's Gallinule
ad
juv
Common Moorhen
ad
juv
Eurasian Coot

PLATE 27: BLACK-AND-WHITE SHOREBIRDS, COURSER AND TURNSTONE

Eurasian Oystercatcher *Haematopus ostralegus* pm

L 43. W 83. Scarce and irregular passage migrant. Occurs Mar–Oct: spring Mar–mid May; autumn Jul–mid Oct, peaking Aug–early Sep. **Habitat** Coastal on migration, rests on beaches and rocky islets; occasionally at salt lakes and pools. **ID** Large ***black-and-white wader with long red bill and rather short red-pink legs (adult)***. Flight strong and direct; ***conspicuous white wingbar***, white rump and ***terminal black tail-band***. Non-breeding adult has white collar and duller bill tip. Juvenile and immature have duller black upperparts, dark tip to bill, greyish-pink legs and white collar. **Voice** Noisy; common call far-carrying *kleep-kleep* and a disyllabic *pick-pick*.

Black-winged Stilt *Himantopus himantopus* PM, MB, wv

L 37. W 75. Common passage migrant, uncommon winter visitor and breeding visitor. In spring, late Feb–early May peaking late Mar–Apr; light autumn passage Aug–mid Oct; increasingly in winter. Breeds late Mar–mid Aug. **Habitat** Shallow salt and freshwater lakes, brackish pools, marshes and water treatment plants; coastal on migration. **ID** ***Exceptionally long, deep-pink legs and slim black-and-white body***; walks with high, graceful carriage. ***In flight, white with uniform black wings and obvious long trailing legs***. Female has slightly browner mantle and scapulars. Non-breeding adult has dusky head and neck markings. Juvenile and immature brownish above with greyish crown and hindneck, and white trailing edge to wings in flight. **Voice** Noisy in breeding areas; a variable sharp *kek*, high-pitched continuous *kikikikik* or *kee-ack*.

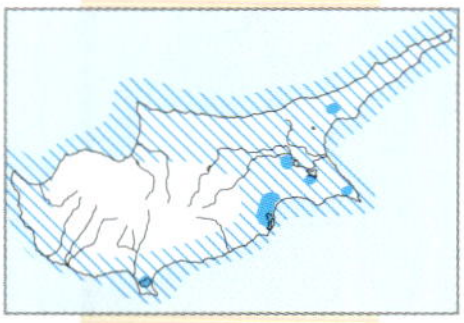

Pied Avocet *Recurvirostra avosetta* pm, wv

L 44. W 79. Uncommon passage migrant and winter visitor. Occurs all months in variable numbers. Mainly Mar–mid May, peaking late Mar–early Apr; fewer Aug–early Dec, occasional in winter, irregular in summer (an opportunist breeder in Middle East). **Habitat** Shallow salt and freshwater lakes, brackish pools and marshes; coastal on migration, often in small groups. **ID** Slender, pied wader with mainly white appearance. ***Black crown and hindneck and long, thin, upcurved bill*** distinctive. Flight stiff-winged and flickering, showing ***white wings with prominent black markings*** and long trailing grey-blue legs. Juvenile tinged brown on black parts, and white in upperparts is mottled buff. Walks steadily and delicately with head down when feeding in shallow water.

Cream-coloured Courser *Cursorius cursor* pm

L 24. W 54. Uncommon passage migrant. Main passage Mar–mid May; rarely in autumn, in Aug; occasionally other months. Alone or in small groups. **Habitat** Sandy or stony semi-desert; short grassy foreshores and ploughed fields; rarely far from coast. **ID** Highly adapted to desert environments. ***Sandy-buff with distinctive black-and-white head markings joining in 'V' on nape, bill short and decurved***. Erect posture; runs quickly with plover-like behaviour and sudden stops, prefers running away rather than flying. Flight rapid with long, slightly rounded wings and legs protruding well beyond tail. Wing pattern distinctive, ***black outer wing above and black underwing contrast sharply with rest of plumage***. Juvenile lacks grey and black on crown; upperparts, head and breast have faint brown spots or irregular dark subterminal lines, and primaries are fringed buff. Sometimes in flocks post-breeding. **Voice** Flight call a short *kwit-kwit*.

Ruddy Turnstone *Arenaria interpres* pm, wv

L 23. W 53. Uncommon passage migrant, rare in winter. Occurs most months. Main passage Mar–early May and late Jul–Sep. **Habitat** Rocky coasts and sandy beaches, foraging on washed-up sea grass; also salt lakes, brackish pools and marshes. **ID** Stocky, ***short-billed and short-legged wader***, distinguished in breeding plumage by ***striking head and breast markings, chestnut shoulders and wing-coverts, and bright orange legs***. Flight strong and direct and shows distinctive pied and chestnut appearance, ***black breast a striking feature***. In winter and juvenile plumages, head and neck much darker, and usually lack chestnut; also juvenile shows bold buffish fringes to most coverts. Walks with a rolling gait, and turns or flips over stones and seaweed for prey. **Voice** Calls short and staccato; when feeding, low *chuk*; in flight, *krytt-te- krytt-te-krytt* or *kritt-it-it*.

ad summer
ad winter
Eurasian Oystercatcher
juv
ad
ad
Black-winged Stilt
ad
Pied Avocet
ad winter
ad
ad summer
Cream-coloured Courser
Ruddy Turnstone

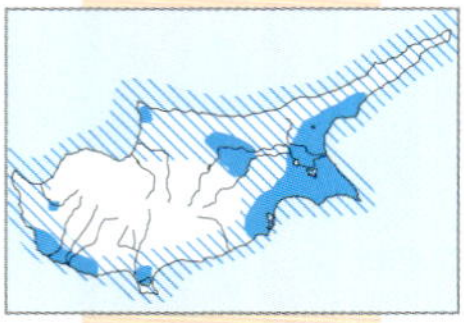

Northern Lapwing *Vanellus vanellus* WV, PM

L 30. Fairly common winter visitor and passage migrant, in variable numbers. Occurs all months, mostly late Aug–early May, with peak numbers Nov–early Feb. Subject to erratic winter influxes, mainly in Jan. **Habitat** Damp fields, salt lakes, marshes, dams and water treatment plants. **ID** Unmistakable with *long, thin, upturned crest and greenish upperparts*. Underparts clean white with broad black breast-band and rich burnt orange undertail. *In rather flappy flight, broad, rounded wings show no wingbar*. In winter, upperparts have narrow pale scaling to feathers. Juvenile lacks tall crest, instead having scruffy, spiky tuft on nape. Often in large flocks outside breeding season. **Voice** Loud, shrill *peeo-vit* uttered in tumbling display flight over breeding grounds.

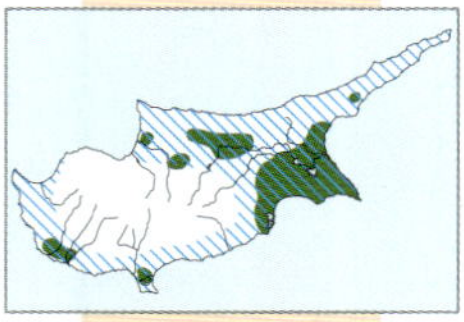

Spur-winged Lapwing *Vanellus spinosus* RB, PM

L 26. W 75. Common breeding resident and passage migrant. Passage Mar–early May and mid Aug–Nov. Breeds mid Mar–Jun, young to mid Sep. Has colonised rapidly since the 1990s. **Habitat** Breeds near run-off areas around sewage and waste-water pools, pig farm slurries and nutrient-rich pools; also fresh and saline marshes and irrigated farmland, with short vegetation. Occurs near coasts on passage. **ID** Elegant, long-legged plover with *black head and underparts, and contrasting white cheeks and neck-sides*. In flight, tricoloured wing pattern: conspicuous white band between black flight feathers and sandy wing-coverts; also broad black tail-band; from below, black belly and flight feathers contrast with white underwing-coverts. Legs black. **Voice** Noisy in breeding areas; alarm call a shrill Oystercatcher-like *dwitt-dwitt* or *kwitt-kwitt*.

Sociable Lapwing *Vanellus gregarius* V

L 29. Vagrant. **ID** Lapwing-sized, with more upright stance, especially when alert. In breeding plumage easily separated by *long white supercilia joining on nape, black crown and chestnut-black belly*. In winter, loses belly patch, becomes mottled on breast, and supercilium is less distinct (but white forehead usually quite prominent). Juvenile, which is browner than adult, has buffish wash to forehead and supercilium, and pale feather edges on upperparts. Legs dark. In flight shows fairly rounded wings, with conspicuous black, white and brownish upperwing pattern, and black band on tip of tail. Typical plover actions on ground, making short runs with head tucked into body, stopping to peck at ground or stand with head erect.

White-tailed Lapwing *Vanellus leucurus* V

L 28. Vagrant. Mainly Mar–early Jun, peaking mid Apr. **Habitat** Marshes, sewage pools and salt lakes. **ID** Slender and graceful, most closely resembles Sociable Lapwing in winter plumage but smaller and has *plain head* (accentuating dark eye), *longer, deep waxy-yellow legs* (which protrude in flight) and, *in flight, all-white tail*. Juvenile paler on neck and breast, dark-mottled on upperparts, with dark cap and faint brown tip to tail. When feeding, tips down so steeply that it almost stands on its head! **Voice** High-pitched *kee-vee-ik*, persistently repeated at nest sites.

Northern Lapwing
ad
ad
Spur-winged Lapwing
ad winter
ad winter
ad summer
Sociable Lapwing
ad
White-tailed Lapwing

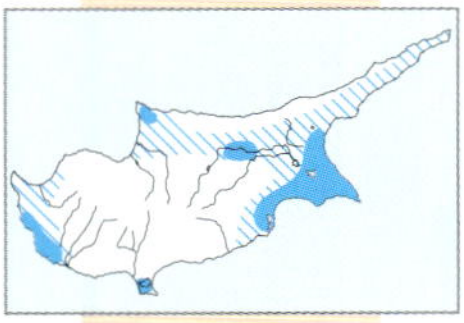

European Golden Plover *Pluvialis apricaria* WV, PM

L 28. W 71. Common winter visitor and passage migrant. Occurs most months: Sep–May, wintering Nov–early Mar, peak numbers late Dec–early Jan. **Habitat** Salt lakes and coastal saltmarshes, on rocky shores and in ploughed fields. **ID** *Resembles Grey Plover, but slightly smaller and daintier with finer bill and narrower wings in faster flight. Upperparts appear brownish with golden hue*. In summer, plumage shows black below. *In flight rather uniform above with faint wingbar, mainly on primaries, and no white on rump or tail; white axillaries and underwing-coverts*. At rest tips of tertials fall well short of wingtip. Larger Grey Plover has greyer appearance, white rump and obvious black axillaries in flight. **Voice** Barely disyllabic melancholy whistle *püyh* or repeated *pyü-pü* (often hard to place). [Alt: Eurasian Golden Plover.]

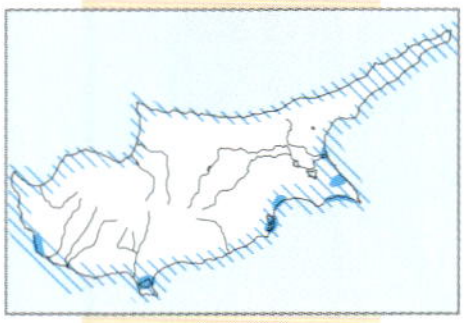

Grey Plover *Pluvialis squatarola* wv, pm

L 29. W 77. Uncommon winter visitor and passage migrant. Usually alone or in small groups. Migration from mid Jul, mainly late Sep–early May, occasionally to Jun. Peak mid Oct–mid Feb. **Habitat** Salt lakes, coastal pools, rocky shores and littoral zones. **ID** Larger than golden plovers with *heavier head and longer, stouter bill. Greyish appearance* lacking obvious yellow or greenish tones above, except faint yellow-buff tinge in juvenile. *Black axillaries diagnostic in flight (all ages)*. Breeding plumage recalls Eurasian Golden, but shows more white on head and nape, and *coarsely speckled black-and-white upperparts*; larger white breast-side patch *without white flanks line. White wingbar and rump* obvious in flight. **Voice** Flight call a mournful trisyllabic whistle, *dee-oo-wee* (second note lower-pitched), often repeated. [Alt: Black-bellied Plover.]

Pacific Golden Plover *Pluvialis fulva* v

L 24. W 66. Vagrant. **ID** Similar to slightly larger European Golden in all plumages. *Best separated by slimmer build, dark underwing and voice*. Compared to European Golden, wings often protrude 1–2cm beyond tail tip, *bill is finer and longer, legs clearly longer*, particularly thighs, making it appear more elegant. *In flight, toes extend beyond tip of tail and wings appear longer and narrower; underwing and axillaries greyish-brown in all plumages* (white in European Golden). Adult in breeding plumage brighter than European Golden, more golden on mantle and scapulars, often with bright white spangles on wing-coverts; non-breeder and immature have more distinct yellow-buff supercilium than European Golden.

Eurasian Dotterel *Charadrius morinellus* pm

L 21. W 60. Scarce passage migrant. Less than annual. In spring, mid Mar–early May, peaking mid Apr. Fewer in autumn, Sep–mid Nov, occasionally Dec–Feb. Usually alone, but sometimes small groups. **Habitat** Ploughed fields, gravel plains and marginal land. **ID** Brownish, dry-country plover; in non-breeding plumage recalls winter European Golden Plover, but smaller with *whitish supercilia meeting in 'V' on nape, and narrow whitish upper-breast-band* in all plumages. Adult breeding female (male duller) has striking *white supercilia and throat contrasting with blackish cap and greyish neck, upper breast bordered white, with chestnut and blackish below*. Adult non-breeding and immature have *brown, scaly upperparts and buff underparts*. *No wingbar in flight*, but white tip to tail and buffish-grey underwing; white shaft on outer primary sometimes rather distinctive. **Voice** Trilling, rather dry Dunlin-like *dryrrr*.

European Golden Plover
1st-winter
Grey Plover
winter
ad summer
winter
ad summer
winter
Pacific Golden Plover
juv
ad ♀ summer
winter
Eurasian Dotterel

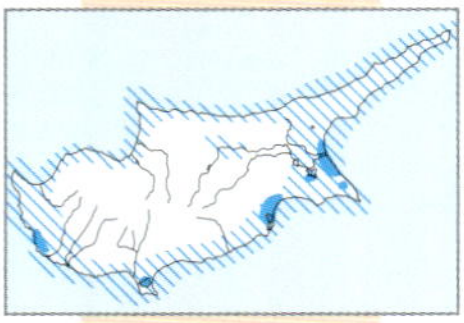

Common Ringed Plover *Charadrius hiaticula* PM, wv

L 19. W 52. Common passage migrant and uncommon winter visitor. In spring Mar–May, peaking early–mid May and in autumn mid Aug–Oct, peaking late Aug–mid Oct. Small numbers in winter and summer. **Habitat** Salt or brackish wetlands, shallow pools, coasts on passage. **ID** Small plover with *orange legs and black-tipped bill, black breast-band and white hind collar.* In flight, *conspicuous white wingbar.* Non-breeding adult has black replaced by dark grey-brown, supercilium and forehead are tinged brown, and bill is all dark. Juvenile similar but paler and duller, often with broken breast-band; upperparts with buff feather fringes. From Little Ringed Plover by white wingbar, lack of pale orbital ring, and *pale supercilium in juvenile. From other plovers by obvious white hind collar and call.* Mainly nominate race occurs around Mediterranean basin, but further east race *tundrae*, with darker upperparts, is more frequent. **Voice** A soft, rising, disyllabic whistle *tooip.*

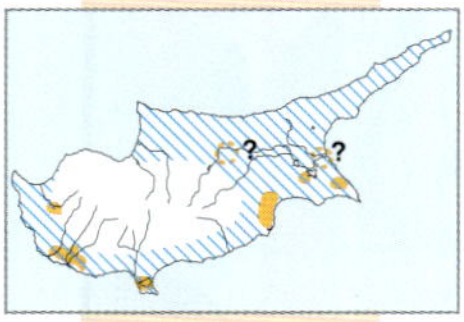

Little Ringed Plover *Charadrius dubius* PM, MB, wv

L 15. W 45. Common passage migrant, scarce winter visitor and migrant breeder. Passage from Feb–early May and late Jun–mid Oct, some overwinter. Breeds Feb–Jun. **Habitat** Fresh and brackish wetlands, dams, shallow pools, water treatment plants and marshes; regular on beaches during autumn passage. **ID** Small, *slim, long-winged plover* with horizontal stance. In breeding plumage similar to Common Ringed Plover, but breast-band narrower and *no white wingbar in flight; yellow orbital ring, dark bill, white line behind black forecrown*, and muddy-coloured or pinkish legs. Adult non-breeder and juvenile have almost plain brown forehead with ill-defined pale patch behind or above eye and duller orbital ring; breast-band broken or absent. Juvenile also has yellow-buff tinge to face and throat. *From all other plovers by call and lack of wingbar.* Mostly alone or in pairs, sometimes small groups. **Voice** Flight and alarm call a loud, plaintive almost monosyllabic *diu*; in wavering display flight, with slow wingbeats, gives *pree-pree-pree* and tern-like *krre-u krre-u.*

Kittlitz's Plover *Charadrius pecuarius* v

L 13. W 42. Vagrant. **Habitat** Muddy wetland margins. **ID** Structure and actions resemble Kentish Plover, but smaller with brownish, *buff-fringed and dark-centred feathers on upperparts, long legs, striking head pattern, and creamy-buff wash to breast and belly.* In adult breeder, white supercilium and black eyestripe both meet at base of hindneck. In non-breeder and juvenile, head markings much duller, lacking black and with white tinged buff. In flight, rather dark with blackish forewing and prominent, but short, white wingbar, most visible on inner primaries; tail pattern similar to Kentish; toes project well beyond tail tip.

Kentish Plover *Charadrius alexandrinus* RB, PM, WV

L 16. W 44. Common resident, passage migrant and winter visitor. Imperceptible spring migration mid Mar–early Apr; obvious passage mid Aug–Oct, often in large numbers. Breeds late Feb–early Aug, frequently double-brooded. **Habitat** Salt lakes and shallow brackish pools, often far inland; also on beaches and rocky shores on passage. **ID** Small, sandy plover with white underparts, rather long, blackish legs, *conspicuous white hind collar and incomplete breast-band*; in flight, a clear white wingbar and *broad white sides to tail.* Adult breeding male has variable *rufous cap, black frontal (forehead) bar, and lateral black breast-patches.* Breeding female, adult non-breeding and juvenile are duller, lack black in plumage, and resemble non-breeding Common Ringed and Little Ringed Plovers, but have rather long, blackish legs, white breast, more white in tail, and different call. By late summer some individuals may bleach to entirely pale sandy-whitish above, with breast-patches often entirely absent. **Voice** Flight call a soft *kip* or *twit*, recalling Little Stint; song a rattled repetition of *tjekke-tjekke* or *jid-id-jid... eer.*

ad summer
winter
Common Ringed Plover
ad summer
winter
Little Ringed Plover
ad summer
imm
ad summer
Kittlitz's Plover
♂ summer
winter
Kentish Plover

PLATE 31: SAND PLOVERS AND CASPIAN PLOVER

Although Lesser Sand Plover has not been recorded in Cyprus, it is a potential vagrant and has been included as it presents an identification challenge because of its similarity to Greater Sand Plover.

Greater Sand Plover *Charadrius leschenaultii* pm, wv

L 24. W 56. Scarce to fairly common passage migrant and winter visitor. Race *columbinus* occurs all months. Migrants arrive mid Jun, mainly Jul–mid Mar. Spring passage Mar–mid Apr/May, when the taller, longer-billed race *scythicus* often also occurs. **Habitat** Flat, rocky intertidal shores in winter, usually alone or in small groups, in south; on passage at salt lakes, brackish pools and beaches. **ID** Medium-sized plover, ***noticeably larger than Common Ringed Plover with longer more pointed bill, larger more angular head and large eye; longer yellowish-green legs, especially thighs, with toes projecting well beyond tail in flight***. Of the two races that occur, *columbinus* is smaller and most resembles the very similar Lesser Sand Plover, which see. Caspian Plover is slimmer with proportionately longer legs, and wings protrude well beyond tail at rest; head more rounded with broader supercilium, underwing dusky, and only a faint white wingbar. **Voice** When flushed a trilling *kyrrr, kirr* or *trrr*.

Lesser Sand Plover *Charadrius mongolus* No records

L 20. W 51. No accepted records, but a potential vagrant. **ID** Very similar to Greater Sand Plover in all plumages, behaviour and shape; lone individuals difficult to identify. Best separated by size (***body slightly larger than Common Ringed Plover; Greater Sand distinctly larger***); ***shorter bill,*** less pointed with more swollen tip (but overlaps with Greater Sand); smaller, more rounded, less angular head; ***shorter, darker, more greyish legs*** (yellowish-green in Greater Sand). At rest, stance often more upright than Greater Sand. Breeding male has ***all-black forehead and mask, and broad reddish-chestnut breast-band*** (Greater Sand never shows all-black forehead, and breast-band usually narrower); female has black on head reduced (like Greater Sand). The '*mongolus* group' (eastern taxa) very difficult to identify except in breeding plumage, when both sexes have prominent white forehead, and male often has narrow black margin to chestnut breast-band. Non-breeder and immature very like Greater Sand Plover, so structural features important. In flight, legs reach to or slightly beyond tail (in Greater Sand legs fairly prominent beyond it); both sand plovers show clear, but variable, white wingbar: in Lesser Sand Plover, this is of more even width (in Greater Sand, often most prominent on inner primaries). **Voice** Quieter than Greater Sand Plover, a short, sharp and less trilling *chitik, chi-chi-chi'*, *chik-tik*; also *kruu-kruit* or *drriiiit*.

Caspian Plover *Charadrius asiaticus* pm

L 19. W 58. Rare spring migrant. Occurs Mar–early Jun, peaking early–mid Apr. **Habitat** Mainly south coast; on ploughed fields, salt flats, grassy plains, semi-desert and brackish pools. **ID** ***Slim, delicate plover*** recalling sand plovers in size and general appearance, but has ***long, attenuated body with wings projecting well beyond tail; proportionately longer legs***, long neck and smaller head with rather fine tapering bill; ***broader white supercilium gives capped appearance***. Male breeding has distinct ***blackish lower border to rufous breast-band***. In flight appears long-winged with faint wingbar visible only on inner primaries, and toes clearly project beyond tail; ***underwing dusky (not white)*** and tail dark with less white at sides and tip. **Voice** Flight call a short, sharp *tyup*, sometimes repeated, and occasionally combined into rapid series of rattling notes *tptptptptp*.

winter
ad summer
columbinus
ad summer
ad summer
scythicus
Greater Sand Plover
ad summer
winter
Lesser Sand Plover
winter
♂ summer
Caspian Plover

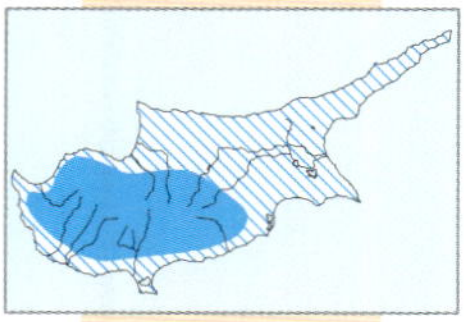

Eurasian Woodcock *Scolopax rusticola* wv

L 34. W 58. Scarce winter visitor, Nov–Mar, with light passage Nov and Mar. **Habitat** Locally common in damp riverine vegetation in upland forest; also wooded areas, riverside brambles, citrus and almond groves, and bushy areas beside streams and dams. **ID** Plumper and larger than snipes; mainly in woodland. Usually seen if flushed; rises silently with zigzagging flight between trees, showing long bill, round body, broad wings, red-brown rump and tail with black subterminal band. On ground note large eyes set far back, barred crown (striped in snipes) and completely barred buff underparts. Often only flushes almost underfoot. **Voice** Distinctive call: one or more deep croaks followed by sharp *twzzip*.

Jack Snipe *Lymnocryptes minimus* pm, wv

L 18 (including bill 4cm). W 40. Uncommon passage migrant and winter visitor, late Sep–early May, peak numbers mid–late Dec and main passage late Mar–Apr. Movements erratic, and may be overlooked due to its secretive nature. **Habitat** Favours marshes, reedbeds, flooded meadows and overgrown pool edges. **ID** Very small with short bill. Well camouflaged and difficult to flush, usually rising silently at c.1m or less, in low, slightly jerky flight, settling almost immediately; two bright yellow lines on upperparts and short, pointed tail distinctive. On ground note green-glossed back, absence of central crown-stripe and dark-striped flanks (not barred). From other snipes by size and short bill. Usually solitary. **Voice** Typically silent when flushed.

Great Snipe *Gallinago media* pm

L 28 (including bill 6cm). W 49. Scarce passage migrant. Mainly in spring, mid Mar–May (exceptionally to mid Jun) peaking mid Apr. Rare mid Sep–Nov. **Habitat** Marshes, damp riverside vegetation, flooded water meadows and lake edges. **ID** Larger, heavier and darker than Common Snipe with shorter bill and closely barred underparts and underwing. When flushed, rises at short distance, flies low and straight without zigzagging; flight heavy with rather slow wingbeats; pot-bellied appearance reminiscent of a small woodcock. In flight, note lack of white belly patch (shown by Common Snipe), dark underwing, white wingbars on greater and median coverts, including primary-coverts, more white on tail corners and absence of white trailing edge to secondaries (conspicuous in Common Snipe). **Voice** When flushed, low-pitched, rather weak, muffled *orrk*.

Common Snipe *Gallinago gallinago* PM, WV

L 26 (including bill 7cm). W 45. Common passage migrant and winter visitor. All months, mostly Aug–May. Spring passage late Feb–May, occasionally Jun, peaking mid Mar–mid Apr. Autumn passage late Jul–Nov, peaking late Sep–early Nov. Winter influxes some years, late Dec–mid Jan. **Habitat** Reedbeds, saltmarshes, water meadows, dams, sewage works, slurry pits and temporary wetlands. **ID** Medium-sized with distinctly yellow-striped head and upperparts, dark-striped breast, barred flanks and very long bill; probes mud with vibrating movements. Often squats low; usually not seen until flushed at 10–15m range; ***rises explosively, immediately uttering several harsh calls, while zigzagging to good height***. In flight, narrow ***white trailing edge to wings and white belly***. **Voice** When flushed, utters a few harsh *ärrtch* notes with slightly rising inflection.

Eurasian Woodcock
Jack Snipe
Great Snipe
Common Snipe

Black-tailed Godwit *Limosa limosa* PM, wv

L 42. W 77. Fairly common passage migrant. Spring mid Feb–May, peaking early–mid Apr, very occasionally Jun–Jul. Rather scarce in autumn, Aug–mid Nov; irregular in winter. **Habitat** Freshwater wetlands, including marshes, dams and sewage works; occasionally salt lakes, brackish pools and coasts on passage. **ID** Large with long legs and *straight bill*. In all plumages easily *distinguished in flight by broad white wingbar, white tail with black terminal band, and trailing legs and feet*. Non-breeder uniform grey, washed ochre on neck and breast in juvenile, with pink-based bill. In summer plumage, both sexes have variable *rusty-orange on head, neck and foreparts with diffuse, dark bars on lower breast, upper belly and flanks*. Female duller, sometimes predominantly greyish. Similar to Bar-tailed Godwit, but slightly larger and more erect due to *longer legs and neck; bill slightly longer and straighter, and totally different upperwing pattern*. **Voice** All calls rather nasal and scolding; alarm call *titi-teev* or, in excitement, *wicka-wicka-wicka*.

Bar-tailed Godwit *Limosa lapponica* pm

L 38. W 75. Scarce passage migrant. Most months, mainly Mar–early Jun and Sep–Nov. **Habitat** Salt lakes, brackish pools and coasts. **ID** Resembles Black-tailed Godwit, especially on ground in non-breeding plumage, but note more compact, less erect appearance with *shorter legs and neck, and slightly upturned bill*. Easily separated in flight by lack of obvious wingbar (and resembles Whimbrel). Non-breeder/winter plumage buffish-grey with dark shaft-streaks above and on breast (almost uniform smooth grey in Black-tailed). Juvenile similar but more buffish with darker upperparts lacking shaft-streaks and breast markings. **Voice** Often silent.

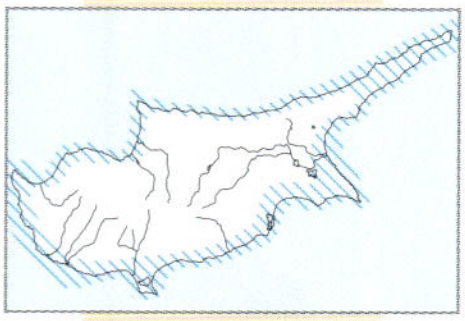

Whimbrel *Numenius phaeopus* pm

L 41. W 83. Uncommon passage migrant. Occurs Mar–mid May, peaking late Mar–mid Apr; and mid Jul–Nov, peaking mid Aug–late Sep. Unusual in winter, although one at Paphos headland most winters since 2011. **Habitat** Rocky shores, salt lakes, brackish pools and beaches with much seagrass. **ID** *Smaller and slightly darker than Eurasian Curlew* with faster wingbeats. *Bill usually shorter and more decurved near tip* (but overlaps with young Eurasian Curlew in length) and head shows *dark crown with pale central stripe and dark eyestripe*. Similar in flight to Eurasian Curlew; body and upperwing usually appearing darker than latter. **Voice** Flight call characteristic, a fast series of whistles, *bi-bi-bi-bi-bi-bi*, quite unlike the soft rising whistle of Eurasian Curlew.

Slender-billed Curlew *Numenius tenuirostris* v

L 39. W 84. Vagrant. **ID** *Much smaller and slimmer than Eurasian Curlew* (smaller even than Whimbrel) with slim neck – never shows 'camel-shaped' neck of Eurasian Curlew. *Shorter, finer, all-dark bill tapering to thin tip* (may show slightly paler base to lower mandible). Plumage 'cleaner' with slightly *greyer upperparts and whiter, more contrasting breast and underparts, showing rounded black spots on flanks and belly-sides in adult* (brown streaks in juvenile). In flight, *dark outer primaries and primary-coverts contrast more with paler inner wing than Whimbrel and most Eurasian Curlews*, but *underwing pure white* (darker in Whimbrel). May now be extinct.

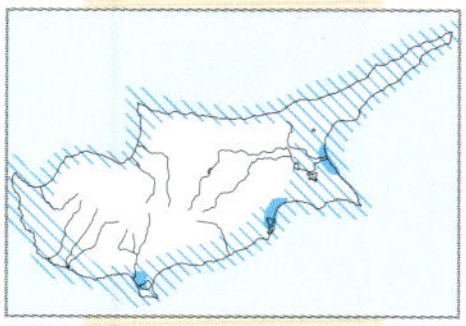

Eurasian Curlew *Numenius arquata* pm, wv

L 55. W 90. Uncommon passage migrant and winter visitor. All months, mainly late Aug–early May, occasionally later, peaking mid Nov–mid Jan and Mar–mid Apr. **Habitat** Marshes, salt lakes, brackish pools and reedbeds; coastal on passage. **ID** Large, streaked, brownish wader with *long, decurved bill*. Male smaller than female, with shorter bill, which in young male overlaps with Whimbrel, but is more evenly decurved. Flight with rather slow, gull-like wingbeats, showing barred tail and white wedge on rump and lower back (like Whimbrel), but *feet project*, unlike Whimbrel. Otherwise separated from Whimbrel by size, *usually longer bill, uniform head pattern and flight call*. **Voice** Flight call a drawn-out, melodic, slowly rising whistle, easily imitated; *cour-leee*.

Black-tailed Godwit
ad summer
winter
winter
ad summer
winter
Bar-tailed Godwit
ad
Whimbrel
ad
Slender-billed Curlew
ad
Eurasian Curlew

Spotted Redshank *Tringa erythropus* pm, wv

L 30. W 64. Uncommon passage migrant, scarce winter visitor. Spring passage Mar–mid May, mostly late Mar–early May; autumn from late Jun, mainly Sep–early Nov. Irregular other months. In breeding plumage, late Apr–Jul. **Habitat** Freshwater pools, water treatment plants, marshes and dams; less often brackish pools and salt lakes. *Often wades or swims while feeding*. **ID** Medium-sized *rather slim wader with red legs*. In winter plumage recalls Common Redshank, but larger with paler grey upperparts, white underparts, and *longer finer bill*, slightly drooping at tip. *Upper mandible wholly black*, lower red with black tip. In breeding plumage *black, finely spotted white*. In flight, all-dark wings, *distinctive white rump extends as wedge up back*, and feet project well beyond tail. **Voice** Flight call a distinct piercing disyllabic *tju-it*.

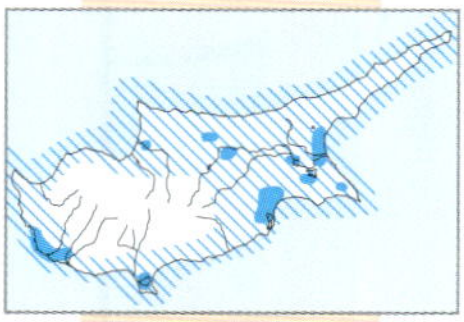

Common Redshank *Tringa totanus* WV, PM

L 28. W 62. Common winter visitor and passage migrant. All months. Passage from mid or late Jun, mainly late Nov–Mar, scarce to mid Jun. **Habitat** Salt lakes, shallow brackish pools, marshes, sewage pools and dams. **ID** Greyish with *bright red legs*, in flight showing *broad white trailing edge to wings and white on rump extending up back*. Winter plumage uniform grey above, paler grey below with ill-defined breast spots. Juvenile has buff-spotted upperparts and all-dark bill. Breeding adult dark-spotted, with red base to bill. Larger and slimmer juvenile Spotted Redshank has greyer upperparts, heavily barred or vermiculated underparts, finer and longer bill, and different wing pattern. **Voice** Calls distinctive; disyllabic *djü-dü*, with stress on first syllable; alarm a persistent *tjü-tjü-tjü....*

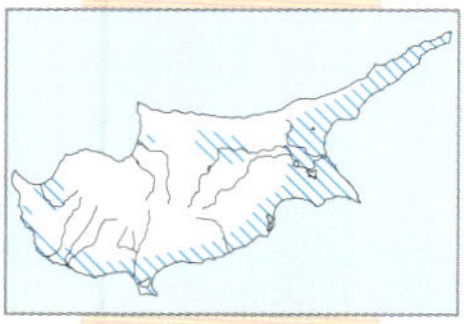

Marsh Sandpiper *Tringa stagnatilis* PM

L 23. W 57. Common passage migrant, Mar–May, peaking Apr, and late Jun–Oct, peaking late Aug–mid Sep. Rare in winter. **Habitat** Most wetlands on passage, particularly marshes, salt lakes, dams and shallow pools. **ID** *Resembles small, slim Common Greenshank with proportionately longer legs and thin, straight, finely pointed bill. Face almost white, usually with supercilium. Flight pattern similar to Greenshank with dark wings, contrasting white tail and wedge on back; dull yellow legs protrude beyond tip of tail*. Flight rapid and action more similar to Wood Sandpiper. Winter plumage rather uniform, paler grey than Greenshank with almost white face and distinctive supercilium. In breeding plumage, becomes markedly black-spotted on head and upperparts; legs usually more yellowish. **Voice** Flight call a clear *djeeu-djeeu* often repeated.

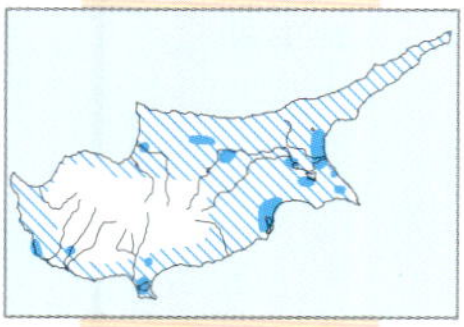

Common Greenshank *Tringa nebularia* PM, wv

L 32. W 69. Fairly common passage migrant; scarce and irregular in winter. Occurs Mar–May, mainly Apr; less common mid Jun–early Nov, peaking mid Aug–mid Sep. **Habitat** Rocky and sandy shores, salt lakes, marshes and brackish pools. **ID** *Larger than Common Redshank with long greenish legs and fairly long, slightly upturned greyish-green bill*. Flight action slow and jerky with long *dark wings contrasting with paler head and neck, and conspicuous white tail and wedge on back*. In winter resembles Marsh Sandpiper but darker grey, and *darker face lacks pale supercilium*. Summer plumage has black spots above and distinctly spotted breast and flanks. Juvenile uniformly patterned with buff-fringed grey-brown upperparts, white underparts, and streaked grey head, neck and breast. Rather active, often runs when feeding in shallow water. **Voice** Flight call a characteristic, shrill, trisyllabic *djiu-djiu-djiu*, with equal stress on all syllables. Marsh Sandpiper's call is similar but thinner and less shrill.

Spotted Redshank
ad summer
winter
winter
ad summer
Common Redshank
winter
ad summer
ad summer
Marsh Sandpiper
winter
ad summer
winter
Common Greenshank

Broad-billed Sandpiper *Calidris falcinellus* pm

L 17. W 35. Scarce passage migrant. Occurs Apr–early Jun, peaking late Apr–mid May; fewer in autumn, mid Jul–Oct. **Habitat** Brackish pools and salt lakes. **ID** Slightly smaller than Dunlin, from which differs by ***longer, broad-based bill with droop near tip*** (head-on, tip also looks swollen), ***shorter yellowish-grey legs and double supercilium***. In summer, upperparts rather dark with white 'V' on mantle, white line on scapulars and ***white underparts with dark-spotted and streaked breast and flanks***. In winter, greyer with less obvious supercilia; then shape and length of bill, leg colour and, if present, dark area on carpals important for identification. In first autumn, resembles adult but streaking below finer and confined to breast. Often has slow-moving and crouching feeding action. In rather erratic flight, appears small, heavy-fronted and, in breeding plumage, dark with narrow wingbar. **Voice** Flight call rather weak with a dry, slightly buzzy character, ascending slightly at end, *brlliid*. [Formerly placed in genus *Limicola*.]

Sanderling *Calidris alba* pm, wv

L 20. W 40. Uncommon passage migrant and scarce winter visitor. Mainly Mar–mid May and mid Jul–Oct. **Habitat** Sandy beaches, salt lakes and shallow pools. **ID** Slightly larger than Dunlin with ***shorter, straighter bill***. In winter, ***very pale plumage with dark mark at bend of wing***. Juvenile more easily confused with other small waders, particularly stints. Note *lack of hind toe*. Summer plumage can be quite ***rusty on head and has prominent breast-band***. Juvenile ***spangled black and white above***, sometimes with buff wash on breast-sides. ***White wingbar noticeable in flight. Often runs fast***, particularly on open shores ahead of wave-wash. **Voice** Usual flight call a loud *plit*.

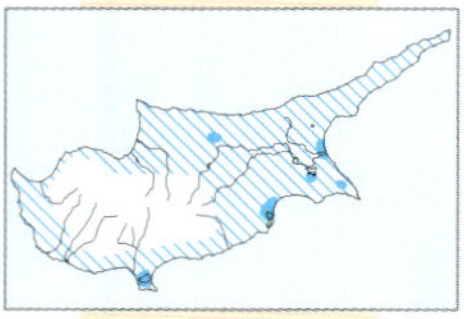

Little Stint *Calidris minuta* PM, WV

L 13. W 30. Very common passage migrant and winter visitor. All months. Spring passage mid Mar–May, mainly mid Apr–mid May; autumn mid Jul–Oct, mainly Sep, with small numbers wintering Nov–Mar. **Habitat** Salt lakes, shallow fresh and brackish pools. **ID** ***Almost always has dark bill and legs***, the latter separating it from Temminck's Stint. In breeding plumage, face, neck, breast and scapulars vary from dull orange to warm buff; but ***always has pale 'V' on mantle and dark centre to crown***. In winter, upperparts become grey, usually with dark shaft-streaks. Juvenile, which looks very white below, has warm rufous tone to upperparts, distinct white 'V' on mantle, diffuse greyish collar and rufous, streaked breast-sides. Feeding action rapid. **Voice** Flight call a short *tip*, *chit* or trill.

Temminck's Stint *Calidris temminckii* PM, wv

L 13. W 30. Fairly common passage migrant, occasional in winter. Passage mid Mar–May, mainly late Apr–early May; and mid Jul–Nov, mainly mid Aug–mid Sep. **Habitat** Brackish and freshwater pools and marshes; less common at salt lakes. **ID** Similar size to Little Stint but body more elongated and ***shorter, yellowish-green legs; white sides to tail prominent in flight***. In summer, mainly grey-buff with ***dark centres to many scapulars***, without rufous-orange tones. In winter, dark scapulars are lost and looks plain buff-grey, but note defined grey breast. Juvenile has narrow buffish fringes to scapulars and coverts with some dark markings on upper scapulars. ***Lacks white 'V' on mantle in all plumages***. Often slightly front-heavy, legs flexed. Sometimes in small groups; often towers high in erratic flight if flushed. **Voice** Flight call a distinctive ringing trill, *tirrrrr*, quite unlike Little Stint.

Baird's Sandpiper *Calidris bairdii* v

L 15. W 38. Vagrant. **ID** Small sandpiper with distinctive shape ***created by long wing projection beyond tail*** and rather short legs. Most likely to occur in region in juvenile plumage, when identified by shape, ***scaly upperparts*** (white fringes to grey-buff feathers), faintly streaked buffish breast, pale spot above lores (at close range), dark legs and bill. In flight, looks long-winged with poorly marked wingbar and poorly defined tail/rump pattern. **Voice** Short, rolling *krru*.

ad summer
winter
winter
Broad-billed Sandpiper
ad summer
winter
Sanderling
ad summer
winter
ad summer
winter
Little Stint
ad moulting
juv
ad spring
Temminck's Stint
Baird's Sandpiper

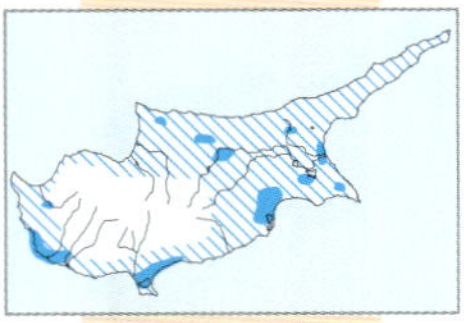

Green Sandpiper *Tringa ochropus* PM, wv

L 23. W 59. Common passage migrant and scarce winter visitor. Occurs mid Feb–early May, mainly mid Mar–mid Apr; and early or mid Jun–Oct, mainly late Jul–early Sep. **Habitat** Prefers non-saline wetlands, including temporary pools, dams, sewage plants and river edges. **ID** Distinguished by call and contrasting black-and-white plumage. Shy, often first seen when flushed, when ***black wings above and below contrast sharply with white belly and white rump***. At rest blackish upperparts and breast contrast with clear white belly and flanks. Juvenile darker, more uniform and buff-spotted, head rather dark with obvious pale eye-ring and short supercilium in front of eye. Often in small flocks. **Voice** Flight call sharp, melodic *dlo-eed-witt-witt*, loud and far-carrying.

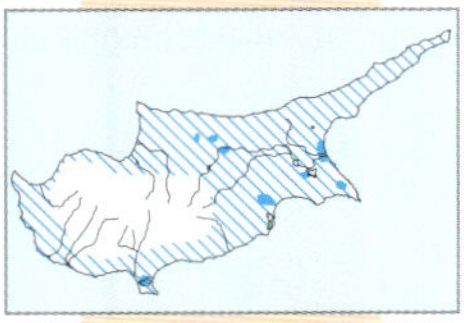

Wood Sandpiper *Tringa glareola* PM

L 20. W 56. Common passage migrant. Occurs Mar–May, mainly Apr–early May, a few to mid Jun; in autumn late Jun–early Dec, mainly late Jul–early Sep. Occasional in winter. **Habitat** Fresh and brackish wetlands, marshes, reedbeds, sewage works, dams and flooded meadows. **ID** Resembles Green Sandpiper but ***upperparts paler, less contrasting and boldly speckled whitish, with conspicuous white supercilium***; also rather long, yellowish-green legs and more elegant appearance. In flight brownish upperparts contrast with white rump; ***differs from Green Sandpiper in whitish (not dark) underwing, longer feet projection*** and call. Juvenile has buff-spotted upperparts, somewhat like young Common Redshank. Occurs alone or in small flocks. **Voice** When flushed, a characteristic, far-carrying series *jiff-iff-iff-iff*.

Terek Sandpiper *Xenus cinereus* V

L 23. W 58. Vagrant; mostly late Mar–May and Aug–Nov. **Habitat** Fresh and brackish pools. **ID** Rather squat, front-heavy wader with short ***orange-yellow legs and long upcurved bill***. Grey head and upperparts ***bordered by dark shoulder bar, carpal patch and primary line. In flight, wings show white trailing edge***, much narrower than that of Common Redshank. In juvenile and winter adult, black shoulder bar faint or absent. ***Frequently bobs rear*** (like Common Sandpiper) and when feeding often runs fast with head lowered. Feeds alone, but may roost in groups. **Voice** Flight call a clear, fluty *tjiy-tjiy* or *dwitt-dwitt*, softer than Common Redshank, sometimes recalling Ruddy Turnstone in character.

Common Sandpiper *Actitis hypoleucos* PM, WV

L 20. W 40. Common passage migrant and fairly common winter visitor. Occurs all months. Spring Mar–May, peaking mid Apr–early May. Spring and autumn passage can overlap in June. In autumn, late Jun–early Oct, peaking late Jul–mid Aug. **Habitat** Streams, pools, dams, sewage works, canals and shorelines; winters on coasts and around dams. Rarely at reedbeds, salt lakes and brackish pools. **ID** Short-legged wader with almost uniform brown upperparts, clean white underparts forming a wedge between wing and brownish breast-sides; rather long tail projects well beyond wingtips at rest; note constantly bobbing rear body. Unique flight action low over water with vibrating, shallow and stiff wingbeats, alternating with short glides. In flight, shows distinct, but rather thin, white wingbar (also obvious from below) and brown rump and tail with pale outer edges. Juvenile has wing-coverts barred buff and dark. Feeds alone, migrates in groups. **Voice** Flight call a characteristic series of descending piping notes, *hee-dee-dee-dee*.

Pectoral Sandpiper *Calidris melanotos* V

L 21. W 42. Vagrant. **Habitat** Marshes and pools with grassy edges. **ID** Long-winged, brown sandpiper, ***larger than Dunlin*** with short, slightly decurved bill and, when not feeding, often a rather long-necked appearance. In all plumages by ***clearly demarcated streaked breast contrasting with white belly, dull yellowish legs and faint wingbar***. **Voice** Loud *kreet* in flight.

winter
ad summer
ad summer
juv
winter
Wood Sandpiper
Green Sandpiper
ad summer
ad
Terek Sandpiper
winter
ad summer
winter
Common Sandpiper
winter
Pectoral Sandpiper

Red Knot *Calidris canutus* pm

L 24. W 50. Rare and irregular passage migrant. Mainly Aug–Nov, less so Apr–May, occasional other months. **Habitat** Brackish and freshwater pools, salt lakes and sewage works. **ID** In non-breeding plumage, identified by combination of stocky build (obviously larger and plumper than Dunlin), pale grey upperparts and straight, rather stout bill (about length of head); in flight, rather long-winged, rump greyish-white and tail grey. Legs grey-green. In breeding plumage, brick-red below with black, white, grey and buff mottling above; from summer Curlew Sandpiper by shorter straight bill, larger size and tail pattern. Juvenile has poorly marked breast; wing-coverts dark subterminal markings and pale tips. **Voice** Short nasal *wut* or *wut wut*, rather quiet.

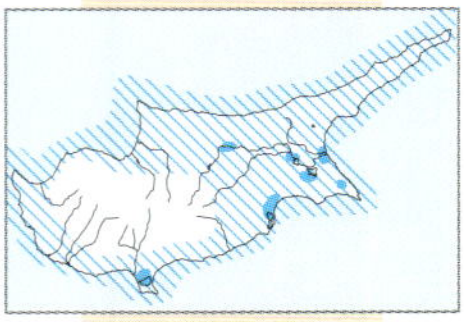

Ruff *Calidris pugnax* PM, wv

L 28 (male); 22 (female). W 56 (male). Common passage migrant, scarce in winter, rare in summer. Migration mid Feb–mid Jun, mainly late Mar–Apr; and mid Jul–Oct, mainly mid Sep. Abundant following seasonal rain and less common in dry autumns. Migrates with other waders and waterfowl. **Habitat** Salt lakes, dams, marshes, shallow brackish and freshwater pools, damp fields and meadows. **ID** Male about size of Common Redshank, but more upright with ***longer neck, smaller head*** and proportionately ***shorter, slightly drooping bill***; often looks humpbacked and pot-bellied. Flight lazy, sometimes interrupted with glides; narrow wingbar and characteristic ***white oval patches at tail base***. Adult in summer plumage varies: upperparts, breast and flanks show mix of black, brown, chestnut, ochre and white, heavily barred or blotched (male may show large ruffs and ear-tufts in spring). Plumage greyer in winter, when ***lores always pale***, face often whitish, and rear belly and undertail always white; a few males show white on head and breast. Bill mostly blackish-brown, but can be yellow or pinkish, tipped dark; ***legs vary from orange-red to greenish-grey. In juvenile, upperparts scaly, blackish-brown distinctly fringed buff, and head, neck and breast yellowish-brown, tinged orange***; legs yellowish-brown or greenish. Often in loose flocks. [Formerly placed in genus *Philomachus*.]

Dunlin *Calidris alpina* PM, WV

L 18. W 38. Common passage migrant and winter visitor. All months, mainly Aug–May. Autumn passage from late Jul, peaking late Nov; spring passage late Feb–mid Mar; occasional to early Jun. With other waders, mainly Little Stint and Kentish Plover. **Habitat** Salt lakes, brackish pools, marshes and other wetlands; sometimes on rocky coasts. **ID** Larger than stints with ***longer, slightly downcurved bill***. Race *alpina* with slightly longer bill (recalling Curlew Sandpiper) predominates; shorter billed *schinzii* is rather scarce, mainly early autumn and late spring. In summer plumage easily identified by ***black belly patch***. In winter, this patch is lost as are rufous tones above; then has grey upperparts with narrow pale fringes to coverts and scapulars, grey breast and white belly. From Curlew Sandpiper by different rump pattern (***white with dark centre***), lack of white supercilium, darker upperparts and breast, and shorter bill and legs. In first autumn, note chestnut on coverts, white 'V' on mantle, and lines of dark spots on flanks below finely streaked breast. **Voice** Reedy *kreep* in flight distinctive.

Curlew Sandpiper *Calidris ferruginea* PM

L 19. W 40. Common passage migrant. In spring earliest from mid–Mar, mainly late Apr–mid Jun; fewer late Jul–Oct, occasional to mid Nov. A few in summer; not reliably reported in winter. **Habitat** Most wetlands, especially salt flats and shallow edges of marshes, dams and sewage pools. **ID** From Dunlin in all plumages by ***white rump, longer, more decurved bill and longer legs***, affording a more elegant appearance. Easily separated in breeding plumage by ***chestnut-red face and underparts*** (often with white feather fringes). In winter plain grey above, white below with light suffusion to breast-sides and ***noticeable white supercilium***; then also distinguished from Dunlin by cleaner appearance with whiter underparts. In first autumn has rather scaly grey-brown upperparts, obvious white supercilium and yellowish-buff wash to breast. Usually in small flocks. **Voice** Trilling *trururip* or almost disyllabic *churrip* in flight.

ad summer
winter
Red Knot
juv
♂ summer
Ruff
winter
♂ summer
ad summer
winter
Dunlin
♂ summer
winter
♀ summer
winter
Curlew Sandpiper

Red-necked Phalarope *Phalaropus lobatus* pm

L 18. W 36. Scarce spring passage migrant, rare in autumn. Occurs Apr–mid May; less frequent late Aug–early Oct. Occasional other months. Usually in ones and twos. **Habitat** Fresh and brackish pools, salt lakes, marshes and dams, mainly in Larnaca and Akrotiri areas. **ID** Small, elegant wader that *swims high on water*, often spinning to disturb food items, taken with *needle-thin, black bill*. Non-breeder has upperparts pale grey with conspicuous *black mask and black on hindcrown*. Breeder has striking head and neck pattern (female brighter than male). Juvenile has similar head pattern, but *upperparts dark brown with prominent ochre bands*. Flight fast and jerky, showing white wingbar; *often settles suddenly on water*. From Red Phalarope by smaller size and very thin black bill. **Voice** Flight call short, sharp *kritt* or *kitt* recalling Sanderling, but quieter and finer.

Red Phalarope *Phalaropus fulicarius* v

L 21. W 42. Vagrant. **ID** Slightly larger and more robust than Red-necked Phalarope, with shorter, thicker neck; *bill thicker, less pointed and sometimes pale (yellowish) at base*. In winter, recalls a very small gull when feeding on water, with *unmarked grey upperparts* (paler, more uniform than Red-necked). First-winter similar to adult, but black on hindneck sometimes present in winter. Behaviour and flight pattern similar to Red-necked Phalarope, but wingbeats less jerky. [Alt: Grey Phalarope.]

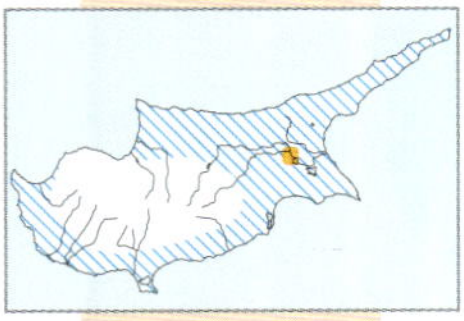

Collared Pratincole *Glareola pratincola* PM, ob

L 25. W 63. Common passage migrant and occasional breeder. Spring Mar–May, mainly mid Apr–early May; fewer Jul–mid Oct. Occasional in summer, and has bred. **Habitat** Salt lakes, brackish pools and marshes; on passage, beaches and ploughed fields. **ID** Distinctive, aerial wader, resembling marsh terns, with *fast graceful flight, long pointed wings, deeply forked tail and short bill*. Usually seen in loose flocks chasing winged insects. On ground, plover-like with quick tripping actions on short legs, often with upright stance, head held high. *Adult has creamy-buff throat bordered black*; tail and wingtips of equal length. In flight, uniform dark olive-brown above (slightly darker flight feathers) with *narrow but distinct white trailing edge to secondaries*, contrasting with white rump and belly; *underwing-coverts reddish-brown*, but often look black. *Some juveniles/worn adults may lack or have only thin white trailing edge to wings*. Black-winged Pratincole has generally darker upperparts, lacks white trailing edge to wings, and has black underwing-coverts. Adult non-breeding/juvenile lack distinct black throat line, and juvenile's brown feathers have pale tips and fringes; outer tail feathers shorter than adult. Sometimes in large flocks. **Voice** Most characteristic call is a tern-like, sharp, chattering *kikki-kirrik*; also a short *check* or *che-keck*.

Oriental Pratincole *Glareola maldivarum* v

L 23. W 63. Vagrant. **ID** Difficult to identify. Closely resembles Collared and Black-winged Pratincoles. Colour of upperparts and *lack of white trailing edge to secondaries (a thin pale fringe may show in fresh-plumaged adult or juvenile) similar to Black-winged, but shares reddish-brown underwing-coverts with Collared Pratincole*. From both in adult plumage by *obviously shorter tail*, best judged when settled, but certainly noticeable in flight. Beware active or suspended moult, or damaged tail, in Collared Pratincole may produce apparently 'short-tailed' individuals.

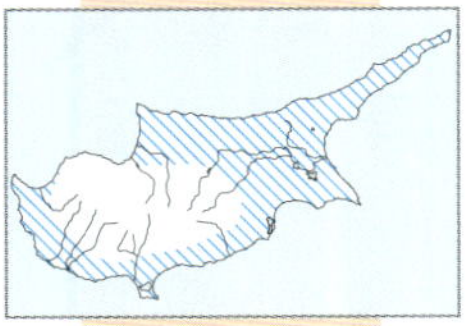

Black-winged Pratincole *Glareola nordmanni* pm

L 24. W 64. Rare passage migrant. Mainly Apr–May and Sep–mid Oct. **Habitat** As Collared Pratincole. **ID** Structure and behaviour as Collared Pratincole and often difficult to distinguish them. Adult darker above than Collared, with no contrast between flight feathers and coverts; *lacks white trailing edge to secondaries, and underwing-coverts are jet-black* (these are reddish-brown in Collared Pratincole but hard to assess, and often appear dark); tail fork of Black-winged shallower. When perched, wings project beyond tail (in Collared approximately equal) and shows less red at base of bill. Non-breeding adult and juvenile resemble corresponding Collared Pratincole plumages but identifiable in flight by wing pattern. **Voice** Call resembles Collared, albeit a little sharper.

Red-necked Phalarope
juv
♀ summer
winter
winter
♀ summer
winter
Red Phalarope
ad
Collared Pratincole
juv
ad
Oriental Pratincole
ad
ad
imm
Black-winged Pratincole

Black-legged Kittiwake *Rissa tridactyla* v

L 40. W 95. Vagrant. **Habitat** Coastal and offshore waters. **ID** Slightly larger than Black-headed Gull. Adult has ***all-black wingtips without white spots***, darker grey back and inner wing but ***whitish, translucent outer wing and black legs***. First-winter recalls smaller Little Gull with ***black 'M' pattern on upperwings***, but head white (except small black spot behind eye) with ***bold, black band on hindneck, white secondaries and slightly forked tail***. First-summer often lacks band on hindneck and sometimes lacks bar on inner wing.

Slender-billed Gull *Chroicocephalus genei* PM, wv

L 43. W 100. Common passage migrant, uncommon winter visitor. Regular late Feb–May exceptionally to Jun, mainly Mar–Apr; in autumn mid Jun–Nov, mainly Aug–Oct. Small numbers in winter. Dependant on water in salt lakes where it feeds by plunging just below surface. **Habitat** Salt lakes, large brackish pools and marshes; coastal on migration. **ID** Larger than Black-headed Gull, ***bill distinctly longer, forehead rather sloping*** with long feathers at base of upper mandible accentuating length of bill. Wing pattern similar to Black-headed, but ***head all white in summer, when breast often has rosy tinge***. Legs longer; when alert ***appears curiously long-necked***. In winter, often has ***small greyish spot on ear-coverts*** (larger, blacker in Black-headed), ***but no vague bar on crown or nape*** (as shown by many Black-headed); ***iris pale in winter*** but dark in many breeders (always dark in Black-headed). ***Dark red bill often appears blackish at distance***. Juvenile and first-winter from Black-headed by pale bill with poorly marked or unmarked tip (always dark in Black-headed) and different head pattern (same as winter adult); pale iris evident during first winter. ***In flight rather hunchbacked, with more protruding neck bulging downwards***. **Voice** Hoarse nasal *yaarr*, deeper and harsher than Black-headed, can recall squeal of distant Water Rail; also abrupt cheery *yap* or *yirp*.

Audouin's Gull *Ichthyaetus audouinii* rb/mb?, pm

L 50. W 125. Breeding resident, dispersive and migratory. Most immatures and some adults probably depart in winter. Light spring migration Mar–early May. Autumn migration and dispersal of breeders Aug–Oct, peaking mid Sep. Small numbers (mainly adults) overwinter. Breeds mid Apr–early Aug; population just 15–30 pairs and apparently declining. **Habitat** Coastal; nests colonially on rocky islands. **ID** Slightly smaller, more delicate than Yellow-legged Gull, with shorter, deeper, ***dark red bill with blackish subterminal band and yellow tip (looks all dark at distance)***, sloping forehead, ***dark olive-grey legs, darker eyes and almost solid black wingtip with just one small white spot near tip***. Juvenile resembles young Lesser Black-backed/Baltic Gull but has ***all-dark uppertail and grey rump with well-defined whitish 'U' on uppertail-coverts; greyer, unstreaked head*** and underparts, less dark ear-coverts and ***grey legs*** (flesh-pink in Lesser Black-backed/Baltic). Compared to most young Caspian Gulls, has darker inner primaries and greater upperwing-coverts, more scalloped upperparts and bolder banded underwing-coverts with more conspicuous mid-wing panel. First-winter retains scalloped effect above (unlike other large gulls); rather white-headed and base of bill olive. Flight more elegant than other larger gulls in region. **Voice** Rather harsh, *grraa(u)*, repeated three or four times.

Common Gull *Larus canus* wv

L 43. W 115. Scarce and irregular winter visitor, mid Nov–mid Apr, mainly late Dec–Mar. Not annual, and usually only immatures. **Habitat** Harbours, beaches, salt lakes and freshwater pools, often with other wintering gulls. **ID** Race *heinei* ('Russian Common Gull'). Adult ***recalls small Caspian Gull, but wings narrower and bill distinctly smaller, thinner and greenish-yellow without red spot***. Black wingtip with prominent white spots, larger than in Caspian. In winter, greenish bill has narrow black band near tip; eyes dark. First-winter *heinei* usually has clean white head and ***bleached-looking wing coverts***. **Voice** Alarm call *klee-uu* with emphasis on last high-pitched note. [Alt: Mew Gull.]

ad
1st-winter
Black-legged Kittiwake
ad winter

ad summer
1st-winter
ad summe
Slender-billed Gull

2nd-summer
juv
ad
Audouin's Gull

ad
1st-winter
ad winter
1st-winter
heinei
Common Gull

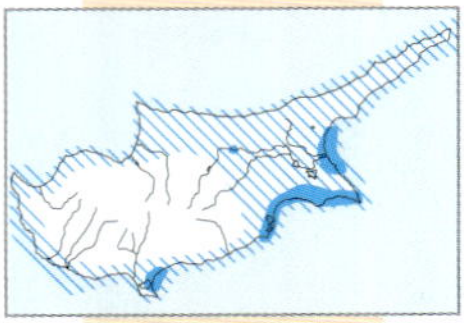

Black-headed Gull *Chroicocephalus ridibundus* WV, PM

L 38. W 93. Very common winter visitor and passage migrant. All months. Some early passage from Jul, mainly late Oct–early Mar; big influx Nov–Feb, when hundreds, even thousands, at favoured sites. A few non-breeders over-summer. **Habitat** Scavenges at ports, harbours and beach restaurants; rests at nearby brackish and freshwater pools; coastal migration. Scarce in south-west. **ID** Medium-small gull, identified (except from Slender-billed Gull) by ***broad white leading edge to primaries, contrasting below with dark grey rest of primaries***. In summer, has ***dark brown hood*** (looks blackish except close to); in winter, head white with black spot on ear-coverts and, often, vague bar from eye over crown and from ear-coverts over nape. Juvenile and first-winter have adult-type primaries; ***from similar Slender-billed Gull by slightly smaller size, shorter bill with darker tip, shorter neck, less bulky breast, more distinct markings on head and always dark eye***. Flight light and buoyant. **Voice** Harsh *krreeea*.

Mediterranean Gull *Ichthyaetus melanocephalus* pm, wv

L 40. W 106. Uncommon passage migrant and scarce winter visitor, late Oct–mid Apr, occasionally earlier, peaking late Nov–mid Dec; and mid Feb–Mar, occasionally to early Jun. **Habitat** Scavenges at ports, harbours and beach restaurants, often with Black-headed Gulls. **ID** Size between Black-headed and Common Gulls. Breeding adult from former by ***all-white primaries, longer thicker bill, paler grey upperparts*** and ***more extensive black hood*** (not dark brown); in winter, head mainly white with ***pronounced dark patch behind eye***. Juvenile and first-winter from similar-aged Common Gull by ***darker, longer, heavier, slightly drooping bill, blacker outer and paler inner primaries and pale mid-wing panel. Fully spread dark outer primaries show white subterminal spots*** (absent in Common Gull, which lacks dark patch behind eye and white eye-crescents). In first-winter, ***back and scapulars much paler than Common Gull***. Subadult has black subterminal spots on wingtip. **Voice** Deep, nasal, far-carrying *ga-u-a* with middle note highest.

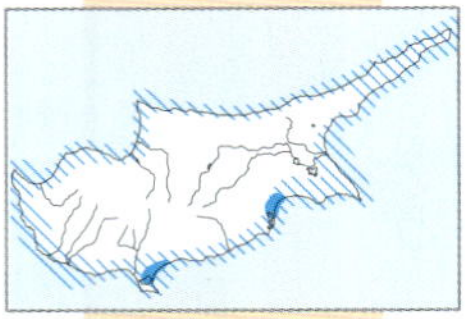

Little Gull *Hydrocoloeus minutus* pm, wv

L 26. W 63. Scarce passage migrant and winter visitor. Major decline since 2005. Occurs late Jul–early Jun, mainly Dec–Apr, with spring peak early Mar–mid Apr. **Habitat** Brackish and freshwater pools, salt lakes and dams; coastal migrant. **ID** ***Much smaller than Black-headed Gull, with tern-like flight***, wingtips fairly rounded though more pointed in juvenile. Adult identified by ***blackish underwing with conspicuous broad white trailing edge***, unmarked pale grey upperwing and, ***in summer, by extensive black hood*** (less extensive, dark brown in Black-headed). In first-winter, underwing pale and ***upperwing has blackish diagonal band***, recalling similar, much larger Black-legged Kittiwake, but ***top of head sooty-grey*** and ***secondaries have narrow dark band*** (Kittiwake has white crown, black band on hindneck and white secondaries). Attains partly black hood as first-summer but has remnants of diagonal band; subadult has black subterminal spots on wingtip. Feeds largely by dipping to water. **Voice** Hard, rather sneezed, marsh tern-like *kjeck-kjeck*.

Franklin's Gull *Leucophaeus pipixcan* v

L 32–36. W 81–93. Vagrant. **ID** Small and compact, size of Black-headed Gull, but no real confusion species in region, though could perhaps be mistaken at rest for Sabine's Gull. Adult has dark grey upperparts, ***white tips to and white band within dark of primaries, red legs and bill, black hood and broken white spectacles***. Immature has dark rear crown and ear-coverts (as winter adult), dark bill, dark grey mantle, and white spectacle behind eye.

White-eyed Gull *Ichthyaetus leucophthalmus* v

L 40. W 108. Vagrant. **ID** A slim, rather dark gull, slightly larger than Black-headed with noticeably ***long, slender bill. White eyelids and dark underwing at all ages***. In adult, bill red with dark tip, ***all-black hood and bib, and grey breast***. Juvenile brownish-grey above with ***greyish head, neck and breast***, white belly and all-dark bill; at distance in flight can be mistaken for a skua.

1st-winter
ad winter
ad summer
ad winter
Black-headed Gull
ad summer
ad winter
1st-winter
juv
Mediterranean Gull
ad summer
ad summer
1st-winter
ad winter
juv
ad summer
ad winter
Little Gull
Franklin's Gull
ad
ad
juv
1st-winter
White-eyed Gull

Pallas's Gull *Ichthyaetus ichthyaetus* pm, wv

L 68. W 158. Scarce to rare passage migrant and winter visitor. Occurs mid Nov–Apr, mainly late Dec–Mar, peaking Mar; occasional in other months. Singles and small parties pass through, mainly juveniles, easily disturbed and often pausing only briefly. **Habitat** Coastal on passage; at dams, salt lakes and large pools. **ID** Adult in summer with *large black hood, large orange-yellow bill with black band near tip* and *white primaries with black band near wingtip*. In winter, large dusky patch behind eye; *white eye-crescents at all ages. At rest appears deep-chested, with long sloping forehead accentuating length of bill*. Readily identified in flight, even at long range, by extensive pale-based primaries (upperwing thus tricoloured). Juvenile and first-winter from other young large gulls by *unmarked white rump and tail with clear-cut band at tip*, unmarked white underparts (though juvenile has grey-mottled breast-band or patches at sides, lost in first-winter), white underwing with extensive black wingtips and often black-tipped coverts forming lines, *pale mid-wing panel above and head pattern*. First-winter from second-winter Caspian Gull by size, *sharp tail-band*, head shape and pattern, dark-mottled hindneck, darker inner primaries and longer bill. Second-winter still has fairly distinctive tail-band. [Alt: Great Black-headed Gull.]

European Herring Gull *Larus argentatus* v

L 54–60. W 123–148. Vagrant. Possibly overlooked and under-recorded. **Habitat** Coastal. **ID** Shorter-winged and dumpier than Caspian and Yellow-legged Gulls with *paler grey upperparts and pink legs* (rarely yellow); head streaked in winter. Grey adult feathers appear in mantle and back from late May of second calendar year. Juvenile and immature have the most extensive translucent inner primaries of any gull in region.

Yellow-legged Gull *Larus michahellis* RB, PM, WV

L 52–58. W 120–140. Very common breeding resident, passage migrant and winter visitor. Autumn passage mid Jul–mid Sep. Winter visitors present Nov–Feb, some immatures remain later; spring passage Mar–Apr. Breeds Feb–Jul. Most adults remain near breeding sites year-round. **Habitat** Breeds on rocky islands and inaccessible cliffs. Outside breeding season mixes with other gulls on coastal waters, dams, salt lakes, sewage pools, rubbish dumps and fishing harbours. **ID** Adult similar to Herring Gull with slightly darker grey mantle, *yellow legs*, squarer head shape with flattish crown, stouter bill and longer wings. *Smaller white mirror on p10 (outermost primary), also usually p9. Greater area of black on outer primaries*. Juvenile and first-winter have pale head with 'shadow' behind eye; heavy black bill and pale back contrasting against dark wings in flight; translucent window on inner primaries barely noticeable. Also pinkish legs and *neat black band on white tail*, often narrower on outer feathers. Adult starts primary moult in Mar, much earlier than other large white-headed gulls. **Voice** Display call deep, rather nasal; also deep *aar* or *wah*.

Armenian Gull *Larus armenicus* WV, pm

L 52–60. W 120–140. Fairly common winter visitor and passage migrant. Occurs Sep–Apr, mainly mid Nov–Mar; small passage mid Sep–early Nov and Mar–mid May. Occasional in summer. Steadily increasing in numbers. **Habitat** Coastal waters, brackish pools and marshes, harbours and dams. **ID** Closest to Yellow-legged Gull, also confusable with Steppe Gull. *Adult generally has dark brownish eye* and can appear like Caspian Gull. Plover-like stance. *Bill typically robust and blunt-ended* with large red gonys spot and *variable, broad, black subterminal band* (may be absent in older birds). Adult averages darker than Yellow-legged Gull, with *orange-yellow legs* (duller in winter); in winter, sparsely streaked hindneck. In flight, upperwing shows large, squared-off black primary area more extensive than Yellow-legged Gull (mirror on outermost primary, p10, and sometimes a small one on p9). Juvenile and older immature similar to Yellow-legged Gull, and distinguished only by structure and call. Adult Armenian Gull starts primary moult later (end Jun) compared to European Herring and Yellow-legged Gulls. **Voice** Distinctive, almost cat-like miaowing call.

ad winter
ad winter
ad summer
1st-year
Pallas's Gull
ad winter
ad summer
1st-winter
European Herring Gull
ad winter
ad summer
1st-winter
Yellow-legged Gull
ad winter
ad summer
1st-winter
Armenian Gull

Great Black-backed Gull *Larus marinus* V

L 71. W 158. Vagrant. **Habitat** Coastal. **ID** Largest of black-backed gulls with conspicuously *heavier bill than other gulls* in region. *Adult has jet-black upperparts* (fading to brown-black), *pink legs* and usually rather dark eyes. Note: Baltic Gull has similarly jet-black upperparts but is much smaller. Flight heavy on broad-based wings. In winter, head and neck white and unstreaked. Juvenile/first-winter similar to smaller Caspian and Yellow-legged Gulls but generally *paler upperparts, more variegated* with broader white tips, bars or notches to coverts and tertials, *dark tail-band narrower* (especially at sides); distinctly dark, fairly uniform underwings, contrasting with whitish body; bill remains black (pale-tipped) until first summer and contrasts distinctly with largely white head. Plain blackish coverts appear in upperparts in second-winter/second-summer, when identification becomes easier.

Baltic Gull *Larus fuscus fuscus* PM, wv

L 53. W 127. Common passage migrant and scarce winter visitor. All months. Spring passage Mar–mid May, peaking late Mar–mid Apr; autumn mid Jul–Oct, mainly mid Aug–early Oct, peaking mid Sep. Irregular in winter and summer. **Habitat** Mainly coastal on migration; occasionally joins gull congregations at Larnaca and Akrotiri; less frequent on inland waters. **ID** Smallest of group, a slim, very long-winged gull. *Adult has jet-black wings and mantle, but can wear greyer, when confusable with Heuglin's Gull especially in winter*, rather sparse streaking on head and neck in winter; *bright yellow legs and one, rarely two, small white mirrors* on wingtip. In flight, wings noticeably long (also evident at rest) and slender. *Little or no contrast in black tone to upperside of primaries.* Flying first-winter shows *evenly coloured dark brown 'hand', secondaries and greater coverts, without translucent inner primaries*, unlike Caspian, Armenian and Yellow-legged Gulls; tertials dark brown, narrowly fringed and tipped buffish-white; tail and rump white, tail base boldly spotted or barred black, with broad, blackish terminal band. *Underwing-coverts dark brown in first-winter and usually lack distinctive pattern* (unlike Caspian and Armenian, which are plain white). As immature plumage progresses, head and underparts turn whitish, and from spring of second calendar year blackish feathers visible on shoulder, back and wing-coverts. **Voice** Calls deep and slightly nasal. [Alt: Lesser Black-backed Gull.]

Continental Lesser Black-backed Gull *Larus fuscus intermedius* No records

L 53. W 127. No accepted records for Cyprus, perhaps overlooked. **Habitat** As other large gulls. **ID** *Adult has dark grey upperparts.* Sparse streaking on white head and neck in winter. *Palest birds doubtfully separable from small Heuglin's Gull, darkest from Baltic Gull. Bright yellow legs and one or two small white mirrors on wingtips.* In flight, wings look long and slender; paler birds have *obvious contrast between black primary tips and rest of wing* (as in Heuglin's); *underside of primaries dark, merging with dark grey subterminal band on secondaries. Long and distinctly pointed primary projection* when perched or on water; rests high on water, *sometimes recalling shape of Common Black-headed Gull, especially in juvenile, when head and bill appear small* and rather pointed. Juvenile similar to, but structurally slimmer than juvenile Yellow-legged Gull. **Voice** Deep and nasal; includes rounded *kyow* and *yug-yug-yug-yug.* [Alt: Lesser Black-backed Gull.]

Great Black-backed Gull

Baltic Gull

Continental Lesser Black-backed Gull

Caspian Gull *Larus cachinnans* WV, pm

L 60–65. W 125–150. Common winter visitor and passage migrant. Occurs mainly late Sep–early Apr, a few to May in some years. Main arrival early–mid Nov, peaking mid Dec–mid Feb. Probably outnumbers Yellow-legged Gull in winter. Movements uncertain, due to its similarity to Yellow-legged Gull. **Habitat** Coastal waters, brackish pools and marshes, rubbish dumps, fish farms, harbours and dams; roosts near fresh and brackish pools. **ID** Long-winged, long-legged, with *long, parallel-sided bill and upright stance. Neat-looking with characteristic 'blank' or 'innocent' facial expression, gently sloping forehead, rounded head and often a dark, 'bullet-hole' eye*, although in some older adults they may be yellow-brown. Adult paler than all likely congeners (except Pallas's Gull). *More white in wingtip than congeners* with large white tongue on inner web of p10 (outermost primary) and often with large all-white tip; some eastern populations have small amount of black within large white wingtip. Grey 'tongues' protrude into more restricted black on primaries. Legs pale greyish-pink/straw; bill has greenish tone to base. Juvenile quickly moults to first-winter, a beautiful gull with clean white head, nestled in 'shawl' around hindneck, and white underparts; pale tips to median and greater coverts form pale borders to finely patterned greater coverts; underwing usually white. Subtle pale line from eye to eye over nape/crown boundary, with 'shadow' behind eye. Note first-winter Heuglin's Gull can look similar to Caspian Gull. Second-winter can appear like an enormous Common Gull. Adult starts primary moult in early Apr. **Voice** Nasal and ringing, sometimes deep, likened to a donkey braying.

Steppe Gull *Larus fuscus barabensis* No records

L 55–65. W 125–150. No accepted records, perhaps overlooked. **Habitat** As other large gulls. **ID** Slightly smaller than Caspian Gull with more rounded head, flattish forehead, medium/long bill, shorter legs and brighter soft parts; *smallish, usually dark eye* – birds that winter in Arabian Gulf often have pale eyes. Usually darker grey than Caspian, slightly paler than Heuglin's Gull. *Grey 'tongues' protrude into black wingtip less than Caspian, but more than Heuglin's; mirrors on p9 and p10 (longest primaries). Underside of flight feathers grey against white coverts* (unlike most Caspian, but similar to Armenian, Yellow-legged and Heuglin's); legs generally deeper yellow in breeding season, but some fleshy. Adult plumage and shape (except head) close to Armenian, with more slender, less blunt-tipped bill, and much less black in wingtip. *May show bill-band.* Winter adult white-headed with weakly streaked hindneck. First-summer has mantle, scapulars and many coverts of adult type. Identification from Caspian and pale Heuglin's Gulls not fully established. Adult starts primary moult in late May. **Voice** Gulping *yah-aah-aah-aah…* and thin, high, drawn-out *peeeer* in winter. [Alt: Lesser Black-backed Gull.]

Heuglin's Gull *Larus fuscus heuglini* pm, wv

L 58–65. W 125–150. Rather uncommon migrant and winter visitor; mid Aug–Jun. Timings difficult to understand but apparently a winter visitor Dec–Mar, with migration mid Feb–mid Mar. Associates with other large gulls in winter. **Habitat** Coasts and brackish pools. **ID** Adult can appear similar in plumage and structure to Continental Lesser Black-backed Gull, but typically larger with more upright stance; more like Steppe Gull but with fiercer face; *one, rarely two white mirrors on wingtip*. Bold streaking on head and neck in winter, to early Apr (to Feb in Steppe Gull). Juvenile can appear similar to Baltic Gull. First-winter/first-summer can appear Caspian Gull-like, but often has bicoloured bill, patterned greater coverts, more heavily marked scapulars, and less translucent inner primaries. Adult starts primary moult in late Jun. **Voice** Deep, nasal *gagaga*; usually silent in winter. [Alt: Lesser Black-backed Gull, Siberian Gull.]

ad summer
ad winter
1st-winter
Caspian Gull
ad summer
ad winter
1st-winter
Steppe Gull
ad summer
ad winter
1st-winter
Heuglin's Gull
Caspian Gull
Steppe Gull
Heuglin's Gull

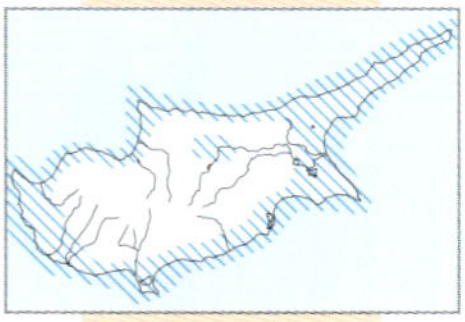

Gull-billed Tern *Gelochelidon nilotica* pm

L 38. W 95. Uncommon passage migrant. Occurs Mar–May, peaking late Mar–Apr and mid Jul–Oct, peaking Aug–early Sep. Occasionally post-breeding groups in summer. Rare in winter. **Habitat** Salt lakes, dams, brackish and fresh pools and farmland; coastal on passage. **ID** Mid-sized tern somewhat resembling Sandwich Tern but ***lacks crest and has shorter, thicker, gull-like all-black bill, shallower forked tail and whitish-grey rump and tail*** (white in Sandwich); lacks contrast shown by Sandwich Tern between dark outer and pale inner primaries above; outer primaries have distinct dark trailing edge. In winter, lacks black cap but ***has variable black 'mask', behind eye***. Juvenile is much less spotted on scapulars and innerwing-coverts than similar Sandwich, and has white crown and ***black eye patch, recalling winter adult Mediterranean Gull***; in first-winter head even whiter. Flight more leisurely and aerobatic than Sandwich. When standing, note more robust build with ***long black legs***. Feeds over fields, marshes, dipping to surface. **Voice** Nasal *ger-wek* with stress on second syllable; alarm fast, agitated, nasal laughing notes, *divdidit*; juvenile call a soft *pre-eep*.

Caspian Tern *Hydroprogne caspia* pm

L 53. W 135. Scarce passage migrant, less than annual. Occurs Mar–Nov, mainly late Mar–early May, peaking mid Apr; less frequent mid Jun–Nov, mainly Aug–early Nov. **Habitat** Coastal migrant; occasionally brackish wetlands, inland pools and dams. **ID** Near size of a large gull, identified by ***large red bill, almost gull-like flight with slow steady wingbeats and by call; distinctive dark primaries below***. Juvenile has orangey bill, more extensive black cap to below eye (unlike adult), weak dark scales on mantle and wing-coverts, primaries dark both above and below; legs pale (black in adult). **Voice** Deep, loud, harsh and deliberate *aark*, recalling Grey Heron, or longer retching *kraa-jak*.

Sandwich Tern *Thalasseus sandvicensis* PM, WV

L 41. W 92. Fairly common passage migrant and winter visitor. Has occurred most months. Mainly Nov–Mar, peaking late Dec–early Jan. Spring passage Feb–Mar, a few to Apr. **Habitat** Coastal, fishing inshore, often on buoys, breakwaters or in mixed gull roosts. **ID** Medium-sized with ***pale grey upperparts and wings (which may look almost white at sea), long slender black bill tipped yellow*** (tip hard to see at distance), long narrow wings, medium-short forked tail and fairly long neck in flight, which is powerful with deep wingbeats. In summer, black cap has ragged crest; in winter, black confined to nape extending forward to eye. Juvenile has all-black bill, dark-scaled mantle and forewing-coverts; first-winter has grey mantle and scapulars, and head pattern like winter adult. **Voice** Loud, grating disyllabic *kerr-rick*; juvenile's call heard into winter, a high-pitched *k-rill*.

Common Tern *Sterna hirundo* pm, ob

L 35. W 80. Scarce passage migrant and occasional breeder. Occurs late Mar–Oct, mainly in Apr, when prospective breeders also arrive. Autumn dispersal and migration mid Aug–Oct. **Habitat** Coastal on migration; salt lakes, fresh and brackish pools near coast. **ID** A medium-sized tern, slightly smaller than Sandwich with (in adult summer) orange-red bill with black tip. Pale grey back and wings with ***dark outer primaries above abruptly demarcated from grey inner primaries; from below translucent innermost primaries (unlike Arctic, which see). Winter adult similar, but has white forehead and mostly black bill***. Juvenile has ***dark grey secondaries*** and dark forewing-band above, with similar translucency and pattern on primaries to adult; bill blackish with red base. **Voice** Sharp *kitt*; *kirri-kirri-kirri*, and a drawn-out *kreee-aeh*.

Arctic Tern *Sterna paradisaea* v

L 38. W 80. Vagrant. **ID** Similar to Common Tern but rounder head, shorter thicker neck, longer tail streamers and more elegant, buoyant flight. ***Bill dark red, usually without black tip; legs shorter. In flight, all primaries appear translucent, revealing a thin, sharp black line to trailing edge of outer primaries***. In summer, greyish underparts with whitish band below black cap. Juvenile/first-winter has ***white secondaries***, diffuse dark bar on leading forewing and white rump.

juv
ad winter
juv
Gull-billed Tern
ad summer
juv
ad winter
juv
Caspian Tern
ad summer
juv
ad winter
Sandwich Tern
ad summer
1st-winter
juv
ad summer
ad summer
Common Tern
ad winter
ad summer
1st-winter
1st-winter
ad summer
Arctic Tern

Little Tern *Sternula albifrons* pm, ob

L 23. W 53. Scarce passage migrant and occasional breeding visitor. Occurs Apr–Sep. Spring migration in mid–late Apr coincides with arrival of breeders, which if successful remain to Jul or early Aug, with dispersal and autumn migration Jul–Sep. **Habitat** Coastal and inland waters; nests on mud and sand banks. **ID** Small with ***fast wingbeats***. Adult in summer is pale grey above, extending in some onto centre of otherwise ***white rump; usually two grey-black leading primaries (with white shafts); white forehead usually extends in point to rear of eye; legs bright yellow to reddish-orange***. In winter, bill black, legs dull grey or brown, usually with some yellow; black on head reduced to band around nape; rump and tail largely grey. Juvenile (dark bill with reddish base) has dark 'U'-shaped markings on pale grey mantle, scapulars and tertials, lost in winter when resembles winter adult; legs greyish-pink to yellow-brown; upperwing shows dark outer primaries becoming progressively paler inwards; leading forewing dark. **Voice** Excited, hoarse grating *kryik* or *pret-pret*.

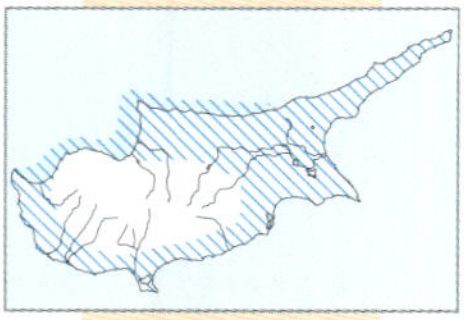

Whiskered Tern *Chlidonias hybrida* PM

L 25. W 73. Common passage migrant. Occurs Mar–Jun, peaking late Apr–early May. More frequent in autumn, Jul–Nov, mainly Aug–Oct, peaking Sep. Spring and autumn migrations overlap, with small numbers present in summer. Rare in winter. **Habitat** Freshwater pools, lakes, marshes, water treatment plants and dams; occasionally salt lakes and irrigated fields. **ID** Slightly larger than White-winged Tern with broader wings, more forked tail (adult), longer legs and ***heavier bill***. Flight buoyant but more *Sterna*-like; dips or dives for food. Adult in summer has ***sooty-grey underparts contrasting with whitish cheek-stripe and underwing***; in winter ***from White-winged by flight action, larger bill, streaked crown and shape of ear-coverts patch***. Juvenile from similar White-winged by head pattern, practically no white collar on hindneck, paler 'saddle' and pronounced blackish and buffish markings on scapulars (contrasting with 'saddle'; scapulars dark in White-winged); upperwing slightly paler with virtually no dark bar on leading forewing; rump and tail usually concolorous pale grey (rump white in White-winged); sometimes has dark smudge at breast-sides, like some adults in winter; absent in White-winged. **Voice** Loud, hoarse *kreck*, almost sneeze-like.

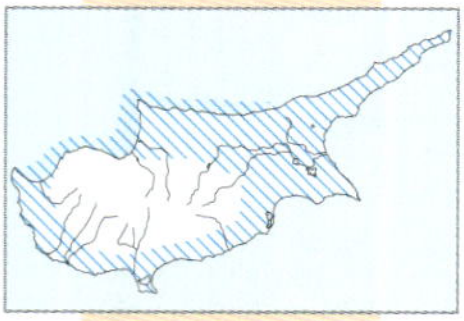

White-winged Tern *Chlidonias leucopterus* PM

L 22. W 65. Common passage migrant. Occurs Mar–Dec. In spring, mid Mar–mid Jun, peaking late Apr–early May. Autumn migration more protracted, late Jun–early Nov, peaking Aug–Sep, occasionally to Dec. The commonest of the three *Chlidonias* terns. **Habitat** Freshwater pools, lakes, marshes, water treatment plants and dams; occasionally salt lakes and irrigated fields; coastal on autumn migration. **ID** Summer adult distinctive; in winter from similar Black Tern by ***lack of dark patch on breast-sides***, whiter rump and tail (pale greyish in Black); a few show black marks on underwing-coverts, making identification easier. ***Juvenile has more contrasting darker 'saddle' and paler upperwing than juvenile Black, and lacks dark breast patches*** and white tips to rear scapulars. Rump and sides of tail whiter, with dark patch on ear-coverts extending further downwards. Moults dark 'saddle' to pale grey in late autumn. **Voice** Harsh, dry *kesch* or *kruek*, less hoarse than Whiskered. [Alt: White-winged Black Tern.]

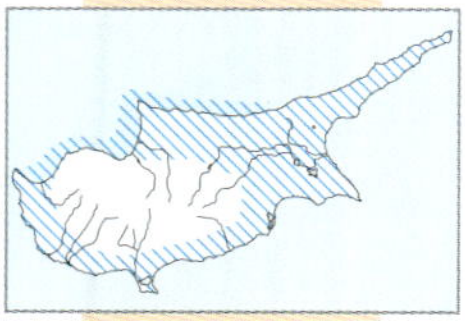

Black Tern *Chlidonias niger* pm

L 23. W 66. Scarce passage migrant. Occurs mid Apr–May/Jun; more frequent Jul–early Nov, peaking mid Aug–early Oct, rarely to early Dec. **Habitat** Freshwater pools, marshes, water treatment plants and dams; occasionally salt lakes; coastal on migration. **ID** Summer adult distinctive; in winter, underparts white with ***dark patch at side of breast***, more pronounced in juvenile; forehead and hind collar white; upperparts greyish, darker and more uniform than White-winged. Juvenile has dark upper mantle and leading upperwing-coverts; scapulars tipped whitish; sides of rump sometimes whitish, rest of rump and tail pale greyish. First-winter recalls adult in winter. **Voice** Short sharp *kjeh* and repeated *kit*, more squeaked than sneezed (as in congeners).

Little Tern

Whiskered Tern

White-winged Tern

Black Tern

Pomarine Skua *Stercorarius pomarinus* v

L 51–56. W 125. Vagrant. **Habitat** Maritime. **ID** Size of large gull, ***heavily built*** with, in adult, diagnostic ***elongated, broad and twisted central tail feathers*** ('spoons'), although these are often lost. Two morphs: dark (scarce) and pale morph, which differ from Arctic Skua in more extensive black cap, darker flanks and blacker breast-band (sometimes absent) and vent. In winter, flanks and tail-coverts barred, elongated tail feathers often short and always blunt-ended. Juvenile best separated from similar Arctic Skua by ***more regular barring on flanks, vent, rump and tail-coverts***. Whitish base to greater primary coverts below often conspicuous, even at long range (frequently absent in Arctic Skua). Bill proportionately larger, heavier at base and more distinctly bicoloured. With experience can be ***identified by larger size, slower wingbeats, broader wings, deeper breast and belly***; also, in juvenile short, blunt central tail feathers (pointed in Arctic Skua). Piratical attacks more direct, less agile than Arctic, frequently chasing large gulls. [Alt: Pomarine Jaeger.]

Arctic Skua *Stercorarius parasiticus* pm

L 46. W 117. Scarce passage migrant. Less than annual. Mainly Mar–mid May, peaking late Mar–mid Apr; fewer late Aug–Oct; occasional in winter. **Habitat** Offshore, very occasionally inland pools. **ID** Streamlined appearance in flight and smaller, less bulky than Pomarine Skua. Adult summer has ***long pointed central tail feathers*** (6–11cm). Two morphs occur (dark and pale), with frequent intermediates. Juvenile variable, from pale to very dark; ground colour generally warmer, more rusty than young Pomarine Skua; juvenile with experience from young Long-tailed Skua *S. longicaudus* (a potential vagrant) by ***broader wings, slightly shorter tail*** (equals width of wing base), ***shorter, pointed central tail feathers*** (longer, blunt-ended in Long-tailed), ***thicker rear body, less black on bill tip and warmer brownish ground colour*** (colder, greyer in Long-tailed), with ***rusty fringes and barring***. Normal flight steady, falcon-like, straight and fast; piratical attacks with sudden twists and turns, harassing birds the size of Sandwich Tern or Common Gull. [Alt: Parasitic Jaeger.]

Pin-tailed Sandgrouse *Pterocles alchata* v

L 35. W 60. Vagrant. Possibly bred prior to 1910. **Habitat** Dry, stony plains and fields. **ID** Large, plump sandgrouse, with ***long tapered tail***. In flight, ***white underparts contrast sharply with black primaries and chestnut-buff breast-band (male) bordered narrowly with black***; female has pale golden-yellow breast-band and ***three black bars***. Male has black throat and eyestripe, and spotted grey-green upperparts; female duller, mottled and barred above. **Voice** Call in flight a distinctive, repeated, slightly falling *arrrh, arrrh* or shorter *arrk-arrk-arrk-*.

Black-bellied Sandgrouse *Pterocles orientalis* wv, FB

L 34. W 72. Rare winter visitor. Weather-related influxes occur some years late Dec–early May, mainly Jan; very occasional summer and autumn. Formerly bred on central plain, where a small relict population may yet survive. **Habitat** Sandy desert areas and gravel plains. **ID** Heavy-bodied sandgrouse. Male has plump ***black belly and prominent black-and-white underwing***. Female has heavily spotted upperparts and breast, sharply demarcated from greyish lower breast and all-black belly. Call highly distinctive. **Voice** Flight call a characteristic, far-carrying ***rolling or bubbling durrrll***, sometimes drawn-out.

pale mcrph
dark morph
ad summer
ad winter
juv
Pomarine Skua
ad summer
ad summer
juv
juv
Arctic Skua
♂
♀
♂
Pin-tailed Sandgrouse
♂
♀
♂
Black-bellied Sandgrouse

Common Wood Pigeon *Columba palumbus* — RB, WV

L 41. Very common breeding resident and fairly common winter visitor. Post-breeding flocks form from mid Jul, joined by winter visitors from mid Sep. Numbers peak late Nov–Jan. Flocks fly between feeding sites and roosts. Breeds Mar–mid Jun. **Habitat** Farmland, foothills and valleys, usually with tall trees, including upland pine forests. **ID** Large stocky pigeon, identified by size, ***bold white band on upperwing in flight*** (all ages) and ***white patches on neck-sides*** (absent in young). Flight slower and heavier than other pigeons, with proportionately longer tail. Gregarious outside breeding season. **Voice** Hoarse cooing, *cu-cooh-cu, coo-coo*, with second note stressed and drawn-out.

Stock Dove *Columba oenas* — wv, pm

L 33. Scarce winter visitor and passage migrant. Occurs late Sep–Apr, mostly mid Oct–mid Mar, with some passage Feb and late Oct–Nov; usually in small numbers. **Habitat** Farmland and open fields. **ID** Easily separated from Common Wood Pigeon by ***absence of white band on upperwing***, also smaller, more compact and shorter-tailed, with quicker wingbeats. From Rock Dove by less white underwing with ***darker flight feathers, pale ashy-grey panel above and lack of bold black bars on secondaries***. Rump grey. Bill pale-tipped (dark in Rock Dove); iris dark. Often in small flocks. **Voice** Monotonous muffled, hollow 'cooing' with emphasis on first syllable, *uu-rur… uu-rur… uu-rur*; also similar single *ur… rer*.

European Turtle Dove *Streptopelia turtur* — PM, MB

L 27. Common passage migrant and breeding visitor. Occurs Mar–mid Nov. Spring passage mainly late Mar–mid Jun, peaking mid Apr–mid May. Autumn passage Aug–Oct, occasionally to mid Nov, peaking in Sep. Breeds Apr–Jul. **Habitat** Wooded areas, farmland, cereal fields and mixed forests at all altitudes. **ID** Small, fast-flying dove with ***shorter tail and more pointed wings*** than Eurasian Collared Dove. Identified by ***darker underwing, well-defined whitish belly, rusty-edged, dark-spotted upperparts*** with blue-grey outer wing panel; also ***contrasting uppertail pattern with clear-cut white corners***, particularly when tail spread on landing. ***Sides of neck show black-and-white-streaked patch in adult***, absent in juvenile. From Laughing Dove by build, jerky wingbeats, scalloped upperparts and tail pattern. Often gregarious. **Voice** Soft, deep purring *roorrrr, roorrrr, roorrrr*, often persistent.

Oriental Turtle Dove *Streptopelia orientalis* — V

L 33. Vagrant. **ID** *S. o. meena* (Rufous Turtle Dove) has occurred. Sometimes treated as a separate species, this race has whitish belly, undertail-coverts and distal part of tail. Resembles European Turtle Dove but ***larger and heavier with broad-based wings, ill-defined dark centres to wing-coverts with narrow buff fringes*** (broader tan fringes in European); ***forehead and crown pale grey, contrasting with browner rest of head***; neck patch larger, ***streaked blue and black*** (usually white and black in European) and ***bare skin around eye rounded*** (larger area bare and lemon-shaped in European). Note also that Oriental Turtle Dove has broad buff fringes to primary-coverts (narrow fringes in European). Juvenile lacks neck patch, is paler brownish on body and wings, otherwise like adult. Flight heavier, straighter, with less jerky wingbeats than European Turtle Dove. **Voice** Alternating grating and clearer cooing notes, *gru-gror, co-co, gru-gror, co-co*, although migrants invariably silent.

Common Wood Pigeon
ad
juv
ad
Stock Dove
ad
European Turtle Dove
ad
ad
Oriental Turtle Dove
meena

Rock Dove *Columba livia* RB

L 33. Breeding resident. Pure Rock Dove populations increasingly diluted by feral pigeons, especially in south, where there are relatively few sightings of 'pure' individuals in recent years. Map includes Feral Pigeon. **Habitat** Quarries, sea cliffs, ravines, mountains; nests in cave or on rock ledge; feral birds also nest under bridges and in towns and cities. **ID** Pale blue-grey pigeon, recalling Stock Dove, with ***two broad black bands on secondaries above; rump white*** or grey; underwing white with dark band at rear; black band on outer tail more pronounced than Stock Dove. Rock Dove is ancestor of familiar Feral Pigeon, which can be blackish, whitish or even reddish; most show black bands on wing, white underwing and contrasting tail pattern. Gregarious, flight very fast and often acrobatic, gliding on lifted wings. **Voice** Cooing *kru-oo-u*, second syllable stressed and higher in pitch.

Eurasian Collared Dove *Streptopelia decaocto* RB, pm?

L 33. Common breeding resident and possible migrant. Numbers peak Mar–early May, late Aug–mid Nov and in midwinter, suggesting migration. Population increasing rapidly, having spread since 1970s through urban and suburban areas. Main nesting period early Mar–Jun; less so Aug–Sep. **Habitat** Around most towns and villages, perched on wires; stubble fields and port grain silos. **ID** Medium-small, sandy-brown dove with narrow black half-collar on hindneck (absent in juvenile). ***From European Turtle Dove by unspotted sandy-brown forewing*** (dark-spotted in Turtle), ***collar, whitish underwing*** (dark in Turtle), plainer uppertail (bold pattern in Turtle), ***absence of well-defined whitish belly***, grey undertail, longer tail and less rapid flight with ***less jerky wingbeats and shorter, more rounded wings***. **Voice** Loud, deep, trisyllabic *coo-cooh-co* with stress on middle note, which is also drawn-out and highest in pitch.

Laughing Dove *Spilopelia senegalensis* rb, v, E/I?

L 26. Uncommon resident and accidental visitor. Some perhaps introduced, but has also colonised naturally. Most vagrants occur in spring and autumn, peaking Mar–Apr and Sep–Oct. Prior to 2013, fewer than five reports of singles; colonies have since formed in several areas and the species is spreading. **Habitat** Towns, villages, gardens and agricultural areas. **ID** Small, dark red-brown to sandy grey-brown dove, with ***black-spotted patch on foreneck and upper breast, unspotted red-brown upperparts and large, blue-grey area on outer wing-coverts***. Uppertail less contrasting than European Turtle Dove. Juvenile lacks patch on foreneck and is duller. Flight like Eurasian Collared Dove but note short rounded wings and long tail. Gregarious. **Voice** Usually five syllables, subdued cooing, with third and fourth notes slightly longer and higher in pitch, *do, do, dooh, dooh, do*. [Alt: Palm Dove.]

Namaqua Dove *Oena capensis* v

L 29 (including 9cm tail). Vagrant. Breeding range expanding in region. **ID** ***Very small, slim dove***, blue-grey or grey-brown with ***long, pointed black central tail feathers***, recalling large budgerigar. In flight, ***black primaries show large red-brown patch***, rump with two transverse black bands. Male has black face and upper breast, which are brownish-grey in female. Juvenile barred black and buff on crown, throat, wing-coverts and back. Flight very fast and direct. Unobtrusive; spends much time on ground. **Voice** Mournful *hu-hu, hu-hu*; also a deep *coo*.

ad
Rock Dove
Feral Pigeon (variations)
ad
Eurasian Collared Dove
ad
display
ad
dark individual
♂
♀
♂
Namaqua Dove
ad
Laughing Dove

Great Spotted Cuckoo *Clamator glandarius* PM, MB

L 40. Common spring passage migrant and breeding visitor. Parasitic, host usually Eurasian Magpie, less often Hooded Crow. Spring passage earliest mid Jan, mainly late Feb–early May. Juveniles from May, common by Jun. Adults depart early May–July. Rare on autumn passage Sep–mid Oct. **Habitat** Farmland, foothills, valleys and plains with scattered trees and bushes. **ID** Large cuckoo with ***slight crest, very long tail, grey upperparts and white-spotted coverts; underparts white***, with buffy throat. Juvenile dark brown above with blackish head, white-spotted coverts and ***distinctive chestnut primaries (above and below)***. Often on ground; flight rather fast, with shallow wingbeats. **Voice** Harsh, chattering *chil, chil, chil, chil*.

Diederik Cuckoo *Chrysococcyx caprius* V

L 18. Vagrant. **ID** Slightly larger than Wryneck with similar shape, and tail often slightly raised. Metallic green male and bronzy female identified by ***white-spotted wings, white supercilium, white crown streak, whitish barring on outer tail feathers*** and ***bold barring on flanks, belly and undertail-coverts***. Juvenile by plain rufous-buff upperparts and barred outer tail feathers. Secretive; appears quite long-tailed in flight. **Voice** Plaintive, loud *pee-pui-pui* (last note often doubled); *psee-psee-psu* (last note lower) or *dee-dee-dee-deederic* (emphasis on last note), often in open from high in tree. [Alt: Didric Cuckoo.]

Common Cuckoo *Cuculus canorus* PM, mb

L 33. Common passage migrant and localised breeding visitor. Occurs Mar–Nov. Common mid Mar–May and those present later are possibly breeders. Scarce late Jul–early Nov. Many heard calling mid Mar–mid Jun. **Habitat** Open country, trees and bushes, clearings in mountain forests. **ID** In flight can resemble Common Kestrel, but head slightly raised and ***wings not lifted above level of body***. Grey above with barred underparts; grey underwing has pale line in centre. Female similar, with warm wash on breast. Rufous morph, in which female and juvenile have brownish-red plumage with noticeable barring on upperparts and breast, relatively common (>25% reportedly of this form in Cyprus). Grey-morph juvenile dark brownish-grey, with pale fringes above and whitish spot on nape. **Voice** Far-carrying song ***unmistakable, kwuk-koo or kwuk, kwuk-koo***, first note highest; female has distinctive, descending, long bubbling call.

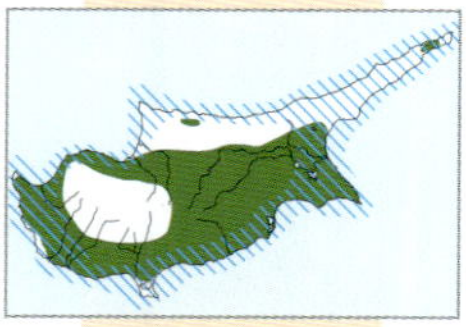

Long-eared Owl *Asio otus* RB, pm, wv?

L 36. W 95. Fairly common breeding resident and uncommon migrant. Shy and secretive. Light passage Mar–Apr and late Sep–early Nov. Some migrants possibly present in winter. Breeds mid Feb–Jun, often in old corvid nest. **Habitat** Breeds in stands of conifers; roosts in large bushes; visible migration at points and headlands. **ID** At roost, muted brown and buff with distinctive ***long ear-tufts*** (invisible when flattened); ***facial disc noticeably warm buff with striking white divide and orange eyes***. Flight jerky; fairly stiff wingbeats with glides on level wings (sometimes slightly raised). From Short-eared Owl by face pattern, ***entirely streaked underparts*** and wing pattern in flight (see Short-eared Owl). Nocturnal. **Voice** Male's territorial call a regularly repeated (c.3-second intervals), hollow *hooo*, like blowing into an empty bottle; alarm call a sharp disyllabic bark, *kwak-kwak*. Young give thin, begging whistle.

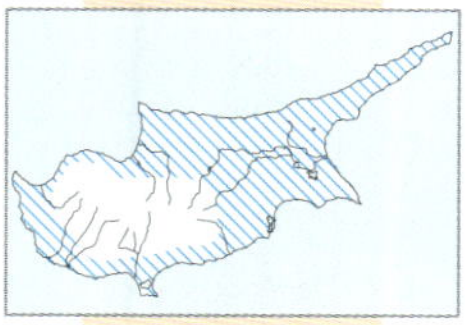

Short-eared Owl *Asio flammeus* pm, wv

L 38. W 102. Uncommon passage migrant, occasionally winter. Mainly mid Feb–early May, peaking Mar–Apr; uncommon mid Sep–Nov and irregular in winter. **Habitat** Coastal on passage; migrants rest in ploughed fields or on sea cliffs. **ID** Often diurnal, flying with slow, ***elegant and high wingbeats*** on long, slender wings, ***raised during glides in shallow 'V'*** (shorter wings with shallower, faster wingbeats in Long-eared); rests on ground. Further differs from Long-eared in strongly buff-spotted upperparts, paler greyish facial disc with striking black surround to ***glaring yellow (not orange) eyes, ear-tufts short***. In flight, by ***yellow-buff base to primaries*** (buff-orange in Long-eared), ***more contrasting bars on flight feathers and tail, white trailing edge to upperwing***, black tips to underwing (not diffusely barred, as Long-eared) and ***dark streaking on breast contrasting with paler belly***. **Voice** Silent on passage.

juv
ad
juv
Great Spotted Cuckoo
♂
♀
Diederik Cuckoo
juv
♂
♀
Common Cuckoo
juv
♂
♀
rufous morph
Short-eared Owl
Long-eared Owl

PLATE 50: BARN OWL AND SMALL OWLS

Western Barn Owl *Tyto alba* RB

L 35. W 89. Fairly common resident. Breeds Mar–Jun; occasionally double-brooded to Sep. Most likely to be seen late Mar–mid Nov; some dispersal in winter. High incidence of road-kills, many blinded by vehicle headlights. **Habitat** Agricultural areas in low-lying areas and foothills, almond and olive groves, and broad river valleys. Nests in abandoned buildings, cliff faces and small caves; rarely above 500m. **ID** Strikingly pale, with *creamy-buff upperparts, white underparts and heart-shaped face with black eyes*. Mainly crepuscular or nocturnal. **Voice** Territorial call of male a clear, vibrant, chilling shriek c.2 seconds long; alarm call in flight a shrill shriek; when disturbed will hiss; young beg with a drawn-out hissing.

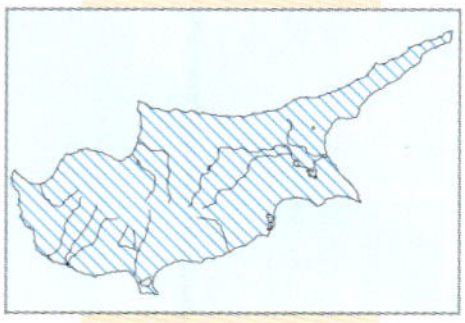

Eurasian Scops Owl *Otus scops* pm, wv?

L 19. Scarce passage migrant. Mainly mid Mar–early May and Sep–Oct; possibly some overwinter; more studies needed. **Habitat** Coastal plains or headlands on passage. **ID** Very difficult to separate from resident Cyprus Scops Owl unless seen well. Typical scops owl with small ear-tufts. Grey-brown with paler face and dark surround to yellow eyes. *Wingtips reach tail tip; whitish tail-bands bordered narrowly black and usually clear-cut* (but often cloaked by closed wings). Upperparts streaked and vermiculated; 'braces' off-white, buff or creamy; underparts streaked, barred and vermiculated, *interspersed by white blotches*. Feathering on legs square-cut short of toe bases. **Voice** Not heard on migration. Clear, soft *whistle pwoo repeated rhythmically*, monotonously so, at 2–3-second intervals.

Cyprus Scops Owl *Otus cyprius* RB, mb?

L 19. Mainly resident; some probably leave in winter. Endemic to Cyprus; now treated as a full species. Breeds Mar–Jul, occasionally earlier, in holes in trees, walls and cliffs, and nest boxes. Usually detected by nocturnal song. Rarely seen by day though sometimes sings then. **Habitat** Common in small towns and villages, lightly wooded areas and forests up to 1,900m. **ID** Compared to Eurasian Scops Owl, *darker and greyer* without rufous tones but larger white spots on hindneck, mantle and cap; also black underparts streaking wider with bars often isolating *distinct white spots* on breast and belly. **Voice** *Best distinguished by diagnostic song*: a monotonous repetition of a two-note phrase *pwoo.... pwoo* (the second note quieter) repeated at 3–4-second intervals.

Little Owl *Athene noctua* RB

L 22. Common resident. Breeds Feb–Jul. **Habitat** Open country, foothills with trees and boulders, abandoned buildings, quarries and cliffs. **ID** Small with *flat-crowned head and long legs*. Race *indigena* is paler than those in Western and Central Europe. *Crown and nape distinctly white-spotted*, with white-blotched upperparts and wings, and *boldly streaked underparts*; eyes yellow, framed white. Flight deeply undulating, alternating rapid flapping with closing of wings (more level in scops owls). Largely crepuscular, but often perches in open by day. If agitated, may bob (pogo) in an upright posture. **Voice** Territorial call a drawn-out wailing *koooah*, faintly falling or rising in pitch at end, also a short *kiu*; alarm call a sharp series of dog-like yapping notes, *kip-kip-kip....*

Western Barn Owl
Eurasian Scops Owl
Cyprus Scops Owl
indigena
Little Owl

PLATE 51: NIGHTJAR AND SWIFTS

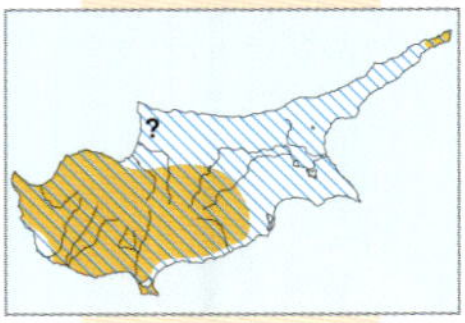

European Nightjar *Caprimulgus europaeus* PM, MB

L 26. Fairly common passage migrant and breeding visitor. Passage from Mar, peaking Apr–early May and late Aug–mid Sep, sometimes to Nov. Breeds mid Apr–early Aug. **Habitat** Mature bushy areas and clearings. Perches on roads at night; feeds over reedbeds, forests and fields. **ID** *Dark-streaked grey crown contrasting with browner cheeks and throat, grey upperparts with broad bands of dark brown and buff-white on scapulars, and rows of whitish to buffy spots on coverts*. Male has white spots in wing and tail. Mostly seen in flight at dawn and dusk; will hunt moths in street lighting or car headlights. In day perches lengthwise along branch or on ground, and if flushed flies short distance on slow, soft wingbeats with long glides on stiff wings recalling cuckoo or falcon, causing small birds to alarm. **Voice** Song at night (rarely in day) long rising and falling *churr*, continuing for minutes, alternating on two pitches and highly ventriloquial. Calls after dusk or pre-dawn, a distinctive, loud throaty *kru-ipp*, or falling, discontented guttural *kworr-kworr*... Display, wing-claps in flight, often giving distinctive *kru-ipp* call.

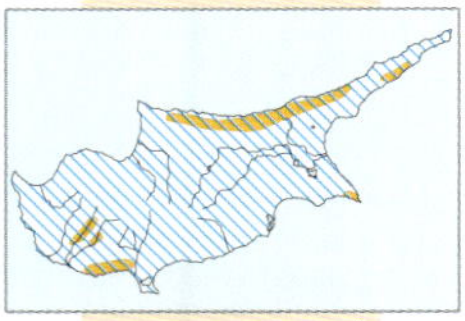

Alpine Swift *Tachymarptis melba* PM, MB

L 21. W 57. Fairly common passage migrant and breeding visitor. Occurs Feb–Nov, occasionally earlier, peaking late Mar–early May. Breeds mid Feb–mid Jun. Dispersal and autumn passage Jul–Nov, mostly at high level, large flocks in Aug. **Habitat** Nests colonially on sea cliffs and steep rock faces; on passage mainly hills in west and north. **ID** Shape and behaviour as Common and Pallid Swifts, but *much larger with markedly slower and deeper, scything wingbeats; white underparts broken by dark breast-band; vent and undertail-coverts brown*. **Voice** Loud, dry chattering trill unlike Common Swift, *trit-it-it-it-itititit-it-it-it*, accelerating then decelerating, rising and falling.

Common Swift *Apus apus* PM, MB

L 16. W 45. Common passage migrant and breeding visitor. Occurs from mid Jan, with spring peak Mar–Apr; less obvious autumn passage, and rare in winter. Breeds Feb–Jul. Thousands of post-breeders over forests Jun–Jul. **Habitat** Anywhere on passage; nests in buildings, old towns, under eaves, occasionally in cliffs. **ID** Race *apus* predominates; *pekinensis* rare on passage. *Almost uniform sooty in fresh plumage*, but brownish, more contrasting in worn plumage and in *pekinensis*, although much depends on light. *Whitish throat patch, sometimes appears absent; lacks contrast above but underparts show inner flight feathers clearly paler than body and wing-coverts*; tail fork deep. Easily confused with Pallid. Wing moult of adults is entirely in winter quarters (unlike Pallid). **Voice** High-pitched ringing *shreee*, variable in pitch; hard to separate from Pallid.

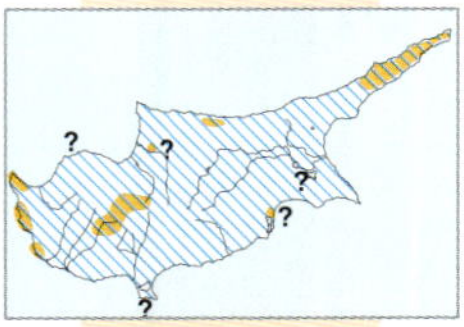

Pallid Swift *Apus pallidus* PM, MB

L 16. W 44. Fairly common passage migrant and uncommon breeding visitor. Occurs from Feb, peaking Apr. Nests mid Mar–Jul, occasionally later. Light autumn passage late Jul–Sep. Small numbers associate with colonies of Common Swift. **Habitat** Nests in sea cliffs; or holes under eaves of older buildings, at all altitudes. **ID** Slightly bulkier than Common Swift but all brown with broader, less tapering outer wing and blunter wingtips, tail fork variable, but shallower with less pointed tips. Brown tones vary, depending on light and wear, and difficult to separate from Common Swift but *head looks broad and flat with larger, more triangular, whitish throat; forehead and lores paler, contrasting with dark eye patch. Faint scaling visible on underparts* close to (rarely seen in Common Swift); *upperparts show darker 'saddle'* slightly contrasting with paler inner flight feathers, head and rump. **Voice** Similar to Common Swift, but deeper, hoarser (tinnier, less ringing; notes less clear-cut) and slightly disyllabic, *sree-er*, ending abruptly.

Little Swift *Apus affinis* pm, ob?

L 12. W 34. Rare migrant and possible breeder, occurring irregularly, mainly Mar–May and Aug–Oct. Reported nest prospecting Jun–Jul. **Habitat** Often migrates with other swifts. **ID** Smallest swift, stout silhouette with *short square-ended tail* (round when spread); prominent *deep white rump-band 'wrapping around' onto rear flanks* (some white visible at all angles). At longer distance can recall House Martin, but note dark underparts, stiff wingbeats and more fluttering flight with short glides. **Voice** Fast, high-pitched, rippling trill *dillillillillill–*.

European Nightjar
♀
♂
Alpine Swift
Common Swift
Little Swift
Pallid Swift

European Roller *Coracias garrulus* PM, MB

L 30. Common passage migrant and breeding visitor. Occurs late Mar–mid Nov. Spring migration variable, mainly late Apr–early May; autumn mid Aug–mid Oct, occasionally to Nov. Breeders arrive mid Apr–early May, occupying nest holes immediately; dispersal from Jul. **Habitat** Open country, usually foothills, broad valleys and agricultural areas; readily perches on wires, poles and dead branches, taking prey on ground. Nests in hole in bank, cliff face, ruin or old tree. **ID** Size of Western Jackdaw. *Turquoise-blue body and most wing-coverts contrasting with blackish flight feathers, chestnut back and deep blue leading forewing above*. Colours often faded in autumn. Juvenile duller and browner with lightly streaked neck and breast. Recalls Western Jackdaw in flight but faster with more vigorous wingbeats. Tail narrow and square-cut. In display flight 'tumbles' from side to side in downward dive (not unlike Northern Lapwing). **Voice** Sonorous, hoarse *rack-rack* recalling Eurasian Magpie; also in display or when agitated a loud piercing *keer-keer-keer* and repeated dry gravelly grating note.

White-throated Kingfisher *Halcyon smyrnensis* v

L 26. Vagrant. Has occurred all months. **Habitat** Coasts, reedbeds, pools and woodland at low elevations. **ID** Large, brightly coloured kingfisher with *enormous red bill, dark chocolate-brown head and belly, large white throat and 'bib', brilliant turquoise-blue upperparts* and black forewing. Flight fast and straight, showing conspicuous white primary patches. Often perches on wires searching for prey on ground. Rather noisy. **Voice** Loud raucous yelping *kril-kril-kril-kril*; also tittering descending song. [Alt: White-breasted Kingfisher, Smyrna Kingfisher.]

Common Kingfisher *Alcedo atthis* PM, WV

L 17. Very common passage migrant and fairly common winter visitor. Occurs mid Jul–mid May. Autumn passage peak Aug–early Nov; winters locally; spring Mar–mid May, with noticeable peak late Mar–mid Apr. **Habitat** Rocky coasts, harbours, marinas, freshwater pools, slow-moving rivers, canals and dams. **ID** Small kingfisher with *brilliant blue and green upperparts, reddish-orange underparts* (often bleached in autumn), but buffish-white throat and neck patch, and long blackish bill (female has mostly red lower mandible). In flight, which is swift, direct and low over water, luminous back and tail obvious. Fairly shy, often inconspicuous when perched, on rock or overhanging branch for long periods before diving for fish; often hovers. **Voice** High-pitched, thin, piercing *cheee* or *tzeee*, mostly in flight.

Pied Kingfisher *Ceryle rudis* pm, wv, ob

L 25. Rare passage migrant, winter visitor and occasional (opportunist) breeder. All months, mainly Nov–Jan and Apr–mid May, peaking mid Apr. Females more common. Migrants have remained on the island for months, wandering between sites. **Habitat** Coasts, dams, fresh or brackish pools, and marshes. Nests in hole in bank. **ID** Large and unmistakable; *the only black-and-white kingfisher in region, frequently seen hovering well above water*, plunging for fish (but also fishes from perch). White underparts have two more or less complete black breast-bands in male, one in female. Black mask, white supercilium, short crest on nape; white-sided tail has black band at tip. Juvenile has greyish breast-band. **Voice** Loud, noisy chattering *chirrick, chirrick, chirrick*.

ad
juv
European Roller
ad
ad ♀
Common Kingfisher
White-throated Kingfisher
ad ♂
Pied Kingfisher

PLATE 53: BEE-EATERS, HOOPOE AND WRYNECK

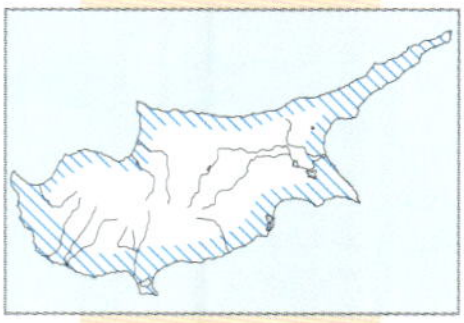

Blue-cheeked Bee-eater *Merops persicus* — pm

L 30. W 48. Scarce spring passage migrant, rare in autumn. Almost annual in variable numbers mid Mar–Jun, mostly late Mar–Apr (one month earlier than European Bee-eater). Fewer in autumn, late Aug–mid Oct. **Habitat** At low elevation often near coasts; reedbeds, irrigated farmland, fertile valleys, headlands and peninsulas. **ID** Larger than European Bee-eater and ***distinctly green or turquoise-green*** with long central tail streamers and strongly ***rusty-red underwing*** framed dark on trailing edge. Juvenile duller and lacks long tail projections. From young European Bee-eater by all-green plumage, including crown and underparts, and rusty-red underwing. Gregarious; hunts insects in flight, often high up. **Voice** Vocal and audible at long range, even if sometimes unseen. Very similar to European Bee-eater, but higher pitched and ***certainly hoarser (throatier)***, the notes sometimes being disyllabic, *prrllip-prrllip* or *prl-rip*.

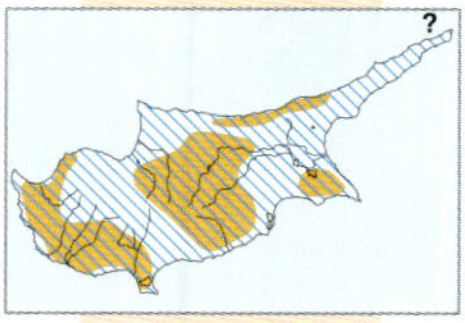

European Bee-eater *Merops apiaster* — PM, mb

L 28. W 46. Very common passage migrant; uncommon breeding visitor. A few from mid Mar, mainly from late Apr, peaking early May, stragglers to mid Jun. Breeds mid Apr–Jul. Family dispersal from Jul; common Aug–early Nov, peaking mid–late Sep, when flocks occur widely. Uncommon, local (or opportunistic) breeder. **Habitat** Migrants feed around farmland and agricultural areas, particularly near beehives for several days; nests colonially in holes excavated in sandy cliffs, riverbanks and quarries, often adjacent to a stream, dam or reedbed. **ID** Easily distinguished from other bee-eaters by ***chestnut crown and back, bright yellow throat contrasting with turquoise-blue underparts*** and, in adult, ***chestnut upperwing-coverts***. In flight shows paler, pinkier underwings than other bee-eaters in region. Juvenile has greenish upperparts, but shows chestnut crown and yellow throat (both paler); central tail feathers mere spikes or lacking. Migrates in vocal flocks, often passing high overhead, sometimes audible but remaining unseen. Hunts insects in flight. **Voice** Similar to Blue-cheeked Bee-eater but softer, lower pitched and more liquid, a far-carrying *prruup*, usually not disyllabic.

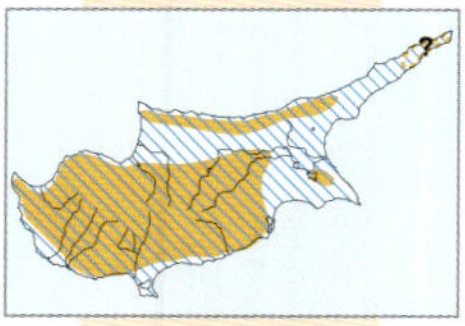

Eurasian Hoopoe *Upupa epops* — PM, MB

L 28. Common passage migrant; fairly common breeding visitor. Spring migration mid Feb–May, peaking late Mar–mid Apr; autumn Aug–early Oct, sometimes to Nov, peaking late Aug–early Sep. Breeds Mar–Jul. Some early autumn migration (or dispersal) from late Jun. Occasionally in winter. **Habitat** Older pine forests, foothills, lightly wooded areas, riverine forests, large gardens and clearings. Hole-nester. **ID** ***Distinctive pinkish-buff bird with bold black and white bars on wings and tail, long black-tipped crest, usually depressed*** (raised on landing), ***long decurved bill; flight with flaps of broad wings recalls Eurasian Jay***. Spends much time on ground; often first noticed when flushed. **Voice** Male's song distinctive, ***repeated hollow poo-poo-poo***; also a dry *terrr* when agitated and strange, thin squeaking and hissing notes in courtship.

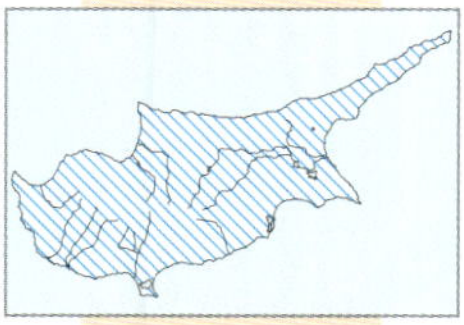

Eurasian Wryneck *Jynx torquilla* — PM, wv

L 16. Common passage migrant; scarce and irregular in winter. Most in spring, from Feb, mainly mid Mar–Apr, peaking late Mar–mid Apr, occasionally May; rather scarce in autumn, mid Aug–Nov, peaking mid Sep. Rare Dec–Jan. **Habitat** Open country with scattered bushes and trees, usually at low elevations. **ID** Inconspicuous, slim, atypical, brown woodpecker with finely vermiculated cryptic plumage. Flight direct, shallowly undulating, reminiscent of large, long-tailed warbler or female Red-backed Shrike. Distinctive features are long tail, ***brown scaling on off-white underparts, dark eyestripe, black central crown streak onto mantle, buff-mottled wings and finely barred yellowish throat***. Often forages on ground. **Voice** Song a loud, monotonous, plaintive *vee-vee-vee-vee...* recalling small falcon.

Blue-cheeked Bee-eater
ad
juv
ad
ad
European Bee-eater
juv
ad
Eurasian Wryneck
Eurasian Hoopoe

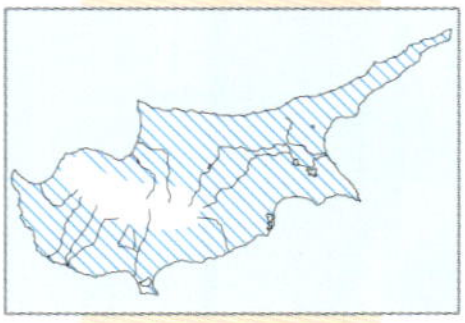

Lesser Kestrel *Falco naumanni* PM, FB

L 28–33. W 63–74. Common spring passage migrant, less common in autumn. Occurs Mar–Jun, mainly late Mar–early May, peaking mid Apr; in autumn Sep–mid Nov, peaking early Oct. Singles have occurred in other months, although winter reports of large foraging flocks are often misidentified Common Kestrels. Former breeder, last confirmed in 1920s. **Habitat** Hunts in variable-sized groups over open country, including over large cereal fields, meadows, ridges and foothills; singles occur anywhere on migration. **ID** Very like Common Kestrel but slightly smaller and slimmer, with narrower wings, more wedge-shaped tail and quicker wingbeats. *Male is unmarked rufous above but greater coverts usually blue-grey. Head ash-grey without moustachial or pale cheeks*. From below, *white underwing contrasts with dark wingtip and creamy-buff body*, both of which have small black spots; in some, underwing-coverts virtually unmarked. Female like Common Kestrel but on average has slightly whiter, less barred flight feathers below, sometimes fewer and finer spots on underwing-coverts and greyer uppertail-coverts. *Female and juvenile can be identified by wingtip formula: p10 (outermost) longer than p8 and clearly longer than p7* or, on close views, by *pale claws* (very rarely black as on Common Kestrel). Hovers less persistently than Common Kestrel, mostly taking insects in flight. Gregarious at breeding sites and on passage. **Voice** Rasping, trisyllabic *chae-chae-chae*, very different from Common Kestrel.

Common Kestrel *Falco tinnunculus* RB, PM, WV?

L 32–38. W 70–78. Very common resident, fairly common passage migrant and possible winter visitor. The most commonly recorded raptor on Cyprus. Autumn passage Sep–Oct, peaking early Oct; less common Mar–early May, peaking early Apr. Migrants often difficult to separate from residents. Breeds Jan–Jun. **Habitat** Most habitats including foothills, farmland, coastal plains and urban areas. Nests in hole or ledge, in tree, cliff or building; will use old nests of other species. **ID** Long, narrow, fairly pointed wings, long, slightly *tapering tail, shallow loose wingbeats, persistent hovering and rufous upperparts, contrasting with darker flight feathers*. Compared to Lesser Kestrel, *male has black spots on back and wing-coverts, lacks blue-grey greater upperwing-coverts*, and has different head pattern and better-marked underwing. Female can approach male in grey on head, tail base and uppertail-coverts. Juvenile paler brown with thin white fringe to greater coverts. Active flight alternates with glides, some soaring and frequent hovering. *At all ages from Lesser Kestrel by black claws and wingtip formula: p10 (outermost primary) shorter than p8 and equal to p7* (useful when soaring at close range outside autumn period of primary moult). Sometimes feeds at high level in large groups following winter rains (cf. Lesser Kestrel). **Voice** Shrill *kee-kee-kee*, often repeated and heard mostly in breeding season.

Merlin *Falco columbarius* pm, wv

L 25–30. W 55–65. Scarce passage migrant and winter visitor. Occurs late Aug–May, mostly Sep–Apr, peaking late Mar and Oct–mid Nov. **Habitat** Open country including farmland, cultivation, plains, marshes and meadows. **ID** *F. c. aesalon* and *F. c. insignis* occur. Immatures and females most likely to be seen. Female larger than male; smallest falcon in region. *Short pointed wings, mid-length tail*, and swift flight on fast wingbeats, interspersed by short glides. *Male by blue-grey upperparts with blackish primaries, broad black tail-band and ill-defined head pattern*. Underparts buffish or whitish with dark streaks, or sometimes rich reddish spotting. Female and juvenile brownish above, creamy below with dark streaks or dense dark spots, with a *diffuse moustachial, barred primaries above and five pale/dark bands of equal width on uppertail*. Hunts usually low over ground with undulating flight, changing direction, followed by direct attack. When perched, wings fall well short of tail tip.

♂
♀
♂
Lesser Kestrel
♂
♀
♂
Common Kestrel
♂
♀
♀
♂
Merlin

PLATE 55: SMALL FALCONS AND PEREGRINE

Red-footed Falcon *Falco vespertinus* PM

L 30. W 73. Very common autumn passage migrant, less common in spring. Occurs mid Aug–early Nov, mainly late Sep–mid Oct; fewer in spring Apr–May, peaking mid–late Apr; rarely other months. Usually feeds in loose groups similar to Lesser Kestrel, but often in greater numbers. **Habitat** Farms, agricultural land, cereal fields, meadows and grassy foothills. Groups rest on ground, including ploughed fields, and often hunt from power lines. **ID** *Male uniform slate-grey with silvery primaries above, red thighs and undertail-coverts, and blackish underwing-coverts. Female has rusty-yellow underparts and head, dark mask and barred tail; yellow-buff underwing has dark trailing edge* (conspicuous in juvenile). First-summer male slate-brown with rufous and slate underparts, but underwing-coverts, flight and most tail feathers as juvenile. In second-winter some adults' central primaries show variegated pattern as moult progresses. Less stocky than Eurasian Hobby in flight with loose, kestrel-like, wingbeats. Gregarious; hunts flying insects, alternating with spells of hovering.

Amur Falcon *Falco amurensis* V

L 26–32. W 65–75. Vagrant. **ID** A small falcon, closely resembling Red-footed Falcon and separated from it by *white underwing-coverts*. Smaller than Eurasian Hobby which it resembles in flight silhouette. Adult male uniform *dark grey with conspicuous pure white underwing-coverts, red thighs and undertail-coverts, grey tail and paler grey underparts than upperparts. Bill and legs bright orange. Adult female white below with warm buff thighs and streaked breast*, boldly barred flanks and *lightly spotted underwing-coverts; dark crown and short moustachial contrast with white cheeks. Juvenile and immature resemble Eurasian Hobby but note white ground colour below and darker crown*. White underwing-coverts appear in male at one year old. Female and juvenile from Eurasian Hobby by barred undertail, smaller moustachial, paler underparts, flight and hovering.

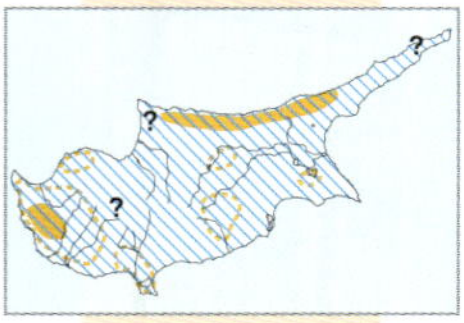

Eurasian Hobby *Falco subbuteo* PM, mb

L 32–36. W 74–92. Common passage migrant and fairly common breeding visitor. Passage late Mar–mid May, peaking mid Apr–early May; more frequent late Aug–Oct (very occasionally to Dec), peaking mid Sep–mid Oct. Breeds mid Apr–Oct. **Habitat** Foothills, farms, wooded valleys, ravines and open country with scattered tall trees, where it also nests. Breeders gather at local reservoirs at dusk to take dragonflies. **ID** *Scythe-like pointed wings* and relatively short tail. Adult has *slate-grey upperparts, uniform tail, prominent moustachial and conspicuous white cheeks*, densely streaked underparts with *red thighs and undertail-coverts*. Juvenile browner above with pale feather fringes, lacks red thighs and undertail-coverts; from juvenile Red-footed Falcon by more distinct breast streaks, darker head with more contrasting mask, unbarred uppertail and underwing pattern. Flight swift and agile; strong steady wingbeats, short fast glides; accelerates when hunting birds, soars when catching insects; rarely hovers (and only briefly). **Voice** Long, rapid *kikikiki*, recalling Common Kestrel but less sharp.

Peregrine Falcon *Falco peregrinus* RB, PM, WV

L 40–52. W 85–120. Fairly common breeding resident, passage migrant and winter visitor. Migrants occur Mar–mid May and late Aug–early Nov. Winter influx late Dec–mid Jan. Breeds Feb–May. **Habitat** Valleys, foothills and coastal farmland; also forests, plains, reedbeds and saltmarshes outside breeding season. Breeds on sea and inland cliffs, gorges and escarpments, occasionally high masts and pylons, at most altitudes. **ID** Large and stocky with relatively short tail and *broad-based, sharply tapering wings, more pointed than Saker and Lanner* when soaring. Adult identified by *black crown and bold moustachial contrasting with white throat and cheeks, barred underparts, white upper breast and uniform underwing. Juvenile has smaller whitish cheeks* (not reaching eye, unlike Saker and Lanner) and uniform underwing (unlike Saker and Lanner). Young of migrant *calidus* race are tricky, showing Saker-like head pattern and large size, but note underwing pattern and wing shape. Local subspecies *brookei* is more compact, like Barbary Falcon, with salmon wash to breast and sometimes rufous wash on nape. Fairly quick, shallow, stiff wingbeats; impressive when hunting, with long, fast stoops. **Voice** Alarm a loud, scolding *aack-ack-ack*.

♀
juv
♂
♀
Red-footed Falcon
♀
♂
♂ 2nd-year
juv
♀
Amur Falcon
ad
juv
Eurasian Hobby
ad
juv
ad
ad
juv
Peregrine Falcon

Eleonora's Falcon *Falco eleonorae* MB, PM

L 39. W 97. Common breeding visitor and passage migrant. Occurs Apr (rarely Mar) to mid Nov, mainly mid Apr–Oct. Spring passage peaks mid–late Apr, autumn from mid Sep. Main departure by late Oct. Breeds May–Sep. **Habitat** Hunts insects over mountain forests and foothills in early summer; also at dams, coasts and wetlands. Breeds in colonies on southern sea cliffs. **ID** ***Long-winged and long-tailed***, recalling Eurasian Hobby but larger. Flight swift and agile, or ***relaxed with slow wingbeats***. Pale morph recalls Eurasian Hobby but has ***darker underparts and dark, unmarked underwing-coverts contrasting with pale-based, unbarred flight feathers***; at distance looks dark below except pale throat and cheeks. Dark morph (25% of population) ***uniform blackish-brown***; from male Red-footed Falcon by size, proportions, flight, underwing pattern, dark primaries above and lack of red thighs. Juvenile (both morphs) paler below than adult pale morph. ***From Eurasian Hobby by dark underwing-coverts contrasting with paler flight feathers, which have dark trailing edge***; also thinner moustachial. Often hunts in flocks, especially at dusk; catches insects in flight, sometimes hovers or stoops. Nestlings coincide with autumn migration of passerines, caught in flight. **Voice** Loud hoarse *kjie-kjie-kjie* when breeding.

Sooty Falcon *Falco concolor* V

L 32–38. W 85. Vagrant. **ID** Larger than Eurasian Hobby and long-winged like Eleonora's Falcon, but tail slightly shorter, with elongated central tail feathers. Adult from dark-morph Eleonora's by ***slaty-grey upperparts with even darker primaries and outer uppertail; blue-grey underparts*** (underwings paler) ***without Eleonora's contrasting underwing pattern***. Female darker than male. Juvenile from similar Eurasian Hobby and pale-morph Eleonora's (adult and juvenile) by ***greyer upperparts with darker wingtips, less clearly streaked underparts, the spot-streaks almost merging on upper breast; lightly marked dusky underwing*** has dark wingtip and trailing edge (Eleonora's has dark coverts contrasting with paler flight feathers; Eurasian Hobby has uniform underwing); ***undertail finely barred except near tip*** (Eurasian Hobby and young Eleonora's have undertail barred to tip). When perched, wingtip reaches tail tip or slightly beyond.

Lanner Falcon *Falco biarmicus* V

L 42–52. W 95–115. Vagrant. **ID** Peregrine-sized, resembling Saker Falcon in plumage and shape; long wings, slightly blunt-ended when soaring; tail relatively long. ***Adult from Saker by barred, greyish upperparts, distinctly barred uppertail***, more contrasting head pattern: ***black forehead band, clear-cut narrow black eyestripe, conspicuous moustachial and spot-bars on flanks***. Race unknown; probably European *feldeggii* which has ***pale rufous rear crown and nape***. Juvenile dark brown above with boldly streaked underparts and rear underwing-coverts; ***closed uppertail is unbarred*** (unlike most Sakers). Contrasting underwing pattern and more densely streaked underparts separate it from juvenile Peregrine and Barbary Falcons. Moderately slow, stiff wingbeats, faster when hunting; stoops or runs down prey; soars with wings level or slightly upturned.

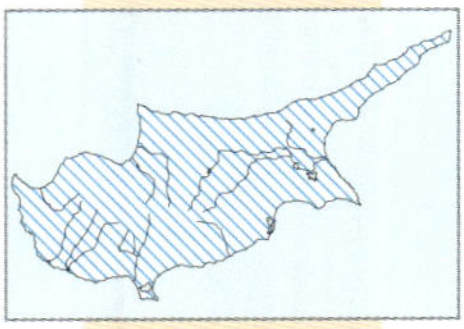

Saker Falcon *Falco cherrug* pm

L 47–55. W 105–125. Scarce passage migrant. Occurs late Feb–May, mainly Mar–early May; in autumn late Aug–Nov, mostly late Sep–Oct, occasionally Dec–Jan. **Habitat** Occurs widely, but most regular at well-watched sites including Akrotiri Peninsula and Cape Greco. **ID** Subspecies *F. c. cherrug* has occurred. Large, heavy-chested with ***creamy crown*** (sometimes just nape), ***unbarred kestrel-like contrast above, less distinctly barred uppertail, poorly developed moustachial and less contrasting head pattern*** (diffuse eyestripe, no dark band on forehead). Whitish supercilium often more conspicuous and belly more spotted, but lacks Lanner's spot-bars on flanks. Saker is greyish above with dark bars, including uppertail; ***best identified by head pattern and size***. Juvenile similar to young Lanner, but outer tail feathers usually conspicuously spotted buff on outer webs (seen well in half-spread tail) and dark stripe behind eye less clear-cut; best by size. When perched, ***wingtip falls short of tail tip*** (unlike most Lanners) and 'trousers' heavier, covering much of tarsus. Slow, flattish wingbeats; when soaring, wings flat or slightly upturned.

Eleonora's Falcon

Sooty Falcon

Lanner Falcon

Saker Falcon

Red-backed Shrike *Lanius collurio* PM, ob

L 18. Common passage migrant, occasionally breeds. Spring migration late Mar–early Jun, mainly mid Apr–May; autumn late Jul–Nov, very common late Aug–mid Oct. **Habitat** Open country, foothills and coasts with scattered trees and bushes. **ID** Adult male unmistakable; female and first-winter warm brown above, tail dark brown, underparts whitish with dense crescentic barring on breast and flanks, also above in immatures. Barring, colour of rump, tail and upperparts separate from Isabelline and Red-tailed Shrikes. Immature similar to Woodchat Shrike (which see). **Voice** Song, sometimes heard on passage, a quiet musical warbling with mimicry. Call a short *shack*, alarm hoarse, hard *keck-keck-keck*.

Isabelline Shrike *Lanius isabellinus* pm

L 16–18. Rare passage migrant. Occurs Mar–early May, peaking mid–late Apr; very rare in autumn. **Habitat** Usually coastal sites, mainly on headlands, favours open areas with perches. **ID** Race *isabellinus*. Pale sandy to sandy-grey above, whitish to creamy- or orange-buff below, with burnt orange flanks. ***Rump and tail foxy- to orange-red***, or dull brown in some females, square-ended or only slightly rounded. Immature has ***upperparts except mantle finely barred*** (crescents on upperparts of Red-backed Shrike have ***dark and pale fringes adjacent***); faintly barred flanks. In female, white at base of primaries often reduced, rarely absent. Adult from Red-tailed Shrike by ***paler upperparts, creamy to warm orange-buff supercilium (never white), pale lores and creamier breast and belly***. Much confusion with Red-tailed in autumn. **Voice** Short, hard, dry staccato ratcheting. [Alt: Daurian Shrike.]

Red-tailed Shrike *Lanius phoenicuroides* pm

L 16–18. Scarce passage migrant. Occurs Mar–mid May, peaking early Apr; rare (or overlooked) in autumn, late Jul–Oct. **Habitat** As Isabelline Shrike. **ID** Sandy grey-brown to rich warm earth-brown above, some almost as rich as Brown Shrike, with contrasting foxy-red rump and tail, latter often ruddier-chestnut than Isabelline Shrike. Spring male from Isabelline by ***rufous crown, white supercilium, darker upperparts and dark lores, also whiter underparts*** with rusty flanks; female and immature also darker above but probably not all immatures safely distinguishable. Intergrades occur. Form *karelini* sandy or sandy-grey above with pale rufous to concolorous sandy-grey crown. **Voice** Calls as Isabelline Shrike. [Alt: Turkestan Shrike.]

Woodchat Shrike *Lanius senator* PM, mb

L 18. Common passage migrant and scarce breeder. More common in spring Mar–mid May, peaking mid Apr; scarce in autumn Aug–Oct, peaking early Sep. Irregular in winter. Breeds mid Apr–early Aug. **Habitat** Open country with scattered bushes, including coasts, valleys and foothills. **ID** Perches prominently; ***rufous-brown crown and nape, and large white shoulder patch on black upperparts*** diagnostic. Female duller than male, with more whitish-buff at base of bill, and rump only slightly greyer than mantle. Juvenile/first-winter resembles juvenile Red-backed Shrike but has ***paler grey-brown upperparts with creamy-white markings on scapulars, wing-coverts, rump and base of primaries*** (pale wing patch noticeable in flight). **Voice** Song comprises varied whistles and trills, often some mimicry; calls *schak-schak* and grating rattles.

Masked Shrike *Lanius nubicus* PM, MB

L 18. Common passage migrant and breeding visitor. Spring passage and arrival of breeders Mar–mid May, peaking early–mid Apr. Breeds late Mar–Jul. Dispersal from late Jun, with migrants very common late Jul–Oct, peaking late Aug–early Sep. Occasional in winter. **Habitat** Migrants favour open country with trees and bushes; in autumn common in foothills. Breeds in mountain forests, occasionally lower altitudes. **ID** Male has ***black upperparts and mask, white face, orange wash on flanks*** (female similar but duller) and distinctive white shoulder patch. ***White bases to primaries and outer tail feathers obvious in flight***. Juvenile similar to juvenile Woodchat but has finer bill, longer tail, paler forehead, colder, ***greyer ground colour to upperparts and rump concolorous with back***. First-winter lacks orange flanks. Perches at mid-levels with ***long tail pumped slowly downwards, or slightly cocked and waved up and down***. Sometimes secretive. **Voice** Harsh, scolding, dry *krrrrrr*. Song rather quiet, a tuneless jumble of notes.

Red-backed Shrike
♀
♂
juvenile
♂
Isabelline Shrike
1st-winter
imm
♂
1st-winter
Red-tailed Shrike
♂
♂
juvenile
Woodchat Shrike
juvenile
Masked Shrike

Lesser Grey Shrike *Lanius minor* PM

L 20. Common passage migrant. Variable numbers in spring mid Mar–May, mainly mid Apr–mid May (most spring migrants pass further east through the Levant); very common in autumn late Jun–mid Nov, peaking late Aug–early Sep, when it is one of the commonest migrant shrikes. **Habitat** Open country with scattered trees and bushes. **ID** Resembles Great Grey Shrike but smaller with ***stouter bill, proportionately longer wings with long primary projection*** and shorter, less graduated tail; ***broader white wing panel*** confined to primaries. ***Adult has extensive black forehead*** without white supercilium, bluish-grey upperparts lacking (sometimes faint) white on shoulders, and ***pinkish-white underparts***. Juvenile/first-winter lacks black forehead, has paler bill, brown-grey upperparts finely barred darker, and paler tips to wing-coverts and flight feathers; underparts creamy-white, sometimes faintly barred. **Voice** Chattering, varied whistles, trills and mimicry.

Steppe Grey Shrike *Lanius pallidirostris* V

L 25. Rare migrant. Late Sep–early Apr, mainly Nov–mid Mar. Individuals often stay for several days or even overwinter. **Habitat** Open country with suitable perches. **ID** Much as Great Grey Shrike, with very pale grey upperparts, white underparts usually blushed pinkish-buff on breast and flanks, ***pale (not black) lores, broad white wingbar not extending to secondaries***; in winter, ***bill horn-coloured or grey***, tipped dark (all black in summer). **Voice** Demonstrative raucous screeches, mechanical throaty whirrs and clicks strung together.

Great Grey Shrike *Lanius excubitor* V

L 25. Vagrant. 'Saharan Grey Shrike' *L. e. elegans* and 'Arabian Grey Shrike' *L. e. aucheri* both occur. **Habitat** Open country with suitable perches. **ID** Resembles Steppe Grey Shrike. Adult varies according to race in shade of grey above and below, and in size of wingbar, but ***lacks white on secondaries***. Lores usually dark with black line above bill; beware confusion with Lesser Grey Shrike, but differs in larger size, wing structure and tail length. Juvenile pale-billed, without black over bill base, unbarred above (unlike juvenile Lesser Grey); from wintering adult Steppe Grey by ochre-buff on breast and wingbars. Race *elegans* is very pale grey above, white below, with large white wingbar and black joining over bill in thin line; *aucheri* slightly darker than *elegans*, with grey wash on flanks, broader black line over bill and smaller area of white in wing. **Voice** Calls include harsh *sheck sheck* often extended into chatter.

Eurasian Golden Oriole *Oriolus oriolus* PM, mb

L 24. Fairly common passage migrant, uncommon breeding visitor. Occurs late Mar–mid Oct. Spring passage late Mar–early Jun, mainly mid Apr–mid May; light autumn passage Aug–mid Oct, mainly Sep. Migrants often sing; easily confused with breeders which are present May–Jul. **Habitat** Orchards, almond groves and fruit trees, from sea level to upper foothills. Breeds mainly in broadleaf woodland valleys of the Troodos and Makheras forests. **ID** Male unmistakable; female and immature male greenish above with olive-brown wings and tail, yellowish-green rump, yellowish-white underparts indistinctly dark-streaked. Appears relatively short-tailed in flight, with woodpecker-like undulations, but strong and fast; often changes direction, tilting and angling wings accordingly. Heard more often than seen. **Voice** Song a loud yodelling *tjoh-wlee-kleeooh*, the last note (often given alone) emphasised and descending; alarm a hoarse, mewing *kra-eik*.

imm
ad summer
Lesser Grey Shrike
imm
ad
Steppe Grey Shrike
ad
aucheri
imm
Great Grey Shrike
ad
elegans
♀
♂
Eurasian Golden Oriole

PLATE 59: CROWS I

Eurasian Jay *Garrulus glandarius* RB

L 33. Common resident. Endemic race *glaszneri* ('Cyprus Jay'). Breeds Apr–Jul. **Habitat** Coniferous and mixed woodland in mountains. In non-breeding season occasionally moves to lower altitudes. **ID** Easily identified, especially in flight, when ***blue wing-coverts, white wing patch and rump contrast with dark flight feathers and tail***. Adult has black moustachial and streaked crown, reddish forehead and smaller bill than other races in Europe and Middle East. Juvenile similar to adult but head pattern less well marked. Flight slow and direct with peculiar jerky flapping of rather broad wings. ***Shy, especially in breeding season, but generally noisy with distinctive call***. Usually forms flocks in autumn when acorns abundant, or when moving to new area. **Voice** Far-carrying, rasping screech, *skaaaak-skaaaak*, also weak mewing *peeay*.

Eurasian Magpie *Pica pica* RB, wv?

L 48. Very common resident with perhaps some arriving in winter. Breeds early Feb–Jul. Main host for Great Spotted Cuckoo. **Habitat** Open country with scattered trees, fertile river valleys and lightly wooded foothills. Builds domed nest, in tree or large thorny bush. **ID** Unmistakable; ***glossy black and white, with long, graduated glossy-green tail; in flight, short, rounded wings with white primaries bordered black***. Flight energetic with irregular flapping wingbeats. Often congregates in large winter flocks (up to 90 together), with so-called 'parliament' on ground. **Voice** Chattering alarm, *chack-chack-chack-chack-*, a weaker *ch-chack* and squealing *keee-uck*.

Yellow-billed Chough *Pyrrhocorax graculus* v

L 38. Vagrant. Has occurred in spring. **Habitat** Mountain peaks. **ID** Larger than Western Jackdaw. ***All black with red legs, slim yellow bill, narrower wings, and proportionately longer, more rounded tail***. Acrobatic flight, often seen soaring over mountains. Feeds on ground on higher rocky hillsides, taking invertebrates. **Voice** Characteristic call a high vibrant trill. [Alt: Alpine Chough.]

Western Jackdaw *Corvus monedula* RB, wv?

L 33. Very common breeding resident and possibly winter visitor. **Habitat** Cliffs, gorges, escarpments and rock faces; also around livestock enclosures. Wanders widely to feed on agricultural land; perches on conifers. Nests colonially on cliffs, less often in holes in trees and buildings. **ID** Small crow with ***grey nape and sides of face, whitish eyes*** and quick movements. Young birds may initially show little grey on head, with iris initially bluish but becoming progressively paler by spring. In flight, compared to other crows, is smaller, has faster wingbeats and distinctly less-'fingered' wings. Often in flocks, frequently with Hooded Crows to feed or at roosts. Walks with quick strides and occasional hops. **Voice** Characteristic, loud, almost metallic *ki-ak* or *kyaar*, harsh drawn-out *jee-aaarah* in alarm; also *ki-ak ggrrrr ggrrrr ggrrr* and *chi-uup*, with other distinctive chuckling notes, often given in conversational chatter.

Rook *Corvus frugilegus* v

L 44–46. Vagrant. Records in all months, but mostly winter. **Habitat** Farmland, reedbeds and rubbish dumps, near tall trees; usually on low ground with other corvids. **ID** All-black purple-glossed crow, adult identifiable by ***long, pale greyish bill with bare grey skin at base, and distinctive calls***. Peaked crown and rather bushy 'trousers'; body feathers generally conceal the tibia, but initially absent in juvenile. ***Tail tip slightly rounded***. Young have less baggy 'trousers' (well into early winter at least), lack pale bill and, until following spring, bare skin at its base; best identified by ***straighter, more pointed bill*** (curved culmen in Carrion Crow) and voice. Other differences subtle: more peaked crown, more measured, deeper wingbeats, wings with longer 'hand' and pinched-in slightly at base. Travels in slow-moving, loose flocks; forages on grassland or fields (often with Jackdaws), walking with bottom-heavy gait. **Voice** Calls measured, with almost languid delivery, less rolling than Hooded and Carrion Crows, deep *kraa* or *raaak*, plus other throaty, glottal notes.

ad
glaszneri
Eurasian Jay
ad
Eurasian Magpie
ad
Yellow-billed Chough
ad
Western Jackdaw
ad
1st-autumn
Rook

PLATE 60: CROWS II AND WAXWING

House Crow *Corvus splendens* V

L 43. Vagrant. The only arrival of this invasive species was at Cape Andreas in Sep 2011; believed to be ship-assisted, remaining almost two months. **ID** Slightly smaller and slimmer than Hooded Crow, with proportionately longer and deeper bill, longer tail and broad-'armed' wings. Flight relatively light and nimble, can recall chough, with slow or fast wingbeats. Readily identified by ***steep forehead, domed crown, grey nape, neck and breast clearly demarcated from black face***, but merging with rest of black plumage. Some races and juveniles show less grey on head or breast. **Voice** Higher pitched than Hooded Crow, a harsh *grehr* or *waaa waaa waaa*; often an even higher *aah - aah*, recalling archetypal crow heard in 'spaghetti westerns'. Can be invasive and drive out most native species. [Alt: Indian House Crow.]

Hooded Crow *Corvus cornix* RB

L 47. Very common breeding resident. Increasing in numbers and expanding range. Breeds Feb–Jun and is secondary host of Great Spotted Cuckoo. **Habitat** Coastal plains and valleys, farmland, foothills, city suburbs and wetlands; not in montane pine forests. Breeds in tall trees, mainly conifers. **ID** Formerly considered a race of Carrion Crow, but now separated. Readily identifiable by ***smoky-grey body (warmer), undertail and underwing-coverts*** with all-black head, breast including bulging bib (with streaks below). Mobs birds of prey. **Voice** Much as Carrion Crow *C. corone*, with similarly rolling or croaking *krarr, krarr* or *aark aark aark*, sometimes hoarse.

Brown-necked Raven *Corvus ruficollis* V

L 50. Vagrant. **ID** Smaller than Northern Raven with ***proportionately longer, slimmer wings, longer head, slimmer bill*** (often held drooping in flight), and ***bronzy-brown sheen on nape and neck*** (can be difficult to see); longish, wedge-shaped tail often shows central feathers protruding beyond rest, whereas Northern Raven is more evenly wedge-shaped. ***At rest wings reach to or beyond tail tip***, in Northern Raven usually fall well short. Juvenile lacks brown on neck. **Voice** Croaking *raark*, less deep than Northern Raven.

Northern Raven *Corvus corax* rb

L 64. Scarce and local breeding resident. Some dispersal or migration in spring and autumn, peaking late Mar and early to mid–Sep. **Habitat** Escarpments and sheer mountain cliffs. Main population in northern Kyrenia Range. Extinct in Troodos since 2008. **ID** A large, powerful corvid similar to Brown-necked Raven (only vagrant to Cyprus) but somewhat larger with broader wings, heavier head and bill, and ***often shows shaggy feathers on throat***. Often soars on flat wings and in breeding season performs tumbling and rolling display flights. **Voice** Loud, far-carrying, ***deep krroak, krroak***, a quieter *kroak* and hollow *klong*.

Bohemian Waxwing *Bombycilla garrulus* V

L 18. Vagrant, last recorded 1966. **ID** Readily identified by soft salmon-grey plumage with ***prominent crest, dark throat and mask, yellow line on wing and yellow-tipped tail***. Undertail-coverts dark rusty-red. Small, red, plastic-like tips on secondaries longest in males and shortest or absent in juveniles. Silhouette rather similar to Common Starling but looks large-headed, slimmer and has more graceful flight. Often in flocks and almost always confiding. **Voice** Call a silvery ringing *sirr*.

ad
House Crow
ad
Hooded Crow
ad
Brown-necked Raven
ad
Northern Raven
ad
Bohemian Waxwing
(not to scale)

Hypocolius *Hypocolius ampelinus* v

L 23. Vagrant. Middle East endemic, greatly sought-after by birders. **Habitat** Fruit-bearing trees and scrub, date-palm groves on passage and in winter. **ID** Recalls slim Great Grey Shrike but has longer tail. ***Soft blue-grey male has black-tipped tail, black mask and black primaries with white tips, prominent in flight***. Female and immature featureless mouse grey-brown, lacking black on head and have only diffuse dark tip to tail. Will fly high with steep climb, often for a distance; note long tail, short wings and rapid wingbeats. **Voice** Mellow flight call, a ***liquid tre-tur-tur,*** the notes run together and the last two are lower pitched. When perched, ***a descending whee-oo***. [Alt: Grey Hypocolius.]

Coal Tit *Periparus ater* RB, v

L 11. Very common breeding resident. Endemic race *cypriotes* ('Cyprus Coal Tit') differs in plumage and song from mainland forms, and is a potential split. Nominate *ater* is a vagrant. Breeds Mar–mid Aug. **Habitat** Uniquely associated with pine forest. In Troodos Mountains to highest peaks, but some wander lower in autumn and winter. Nests low in stump or hole in tree, bank or ground. **ID** ***Black head, white cheeks and diagnostic white nape patch. Small restless tit***, rather dark and distinctly short-tailed with behaviour and some calls recalling Goldcrest. Often in curious, loose parties. *P. a. cypriotes* is darker than mainland forms, rich brown above, ***black on head extending to upper mantle***, and underparts generally very dark with rufous-brown flanks; ***breast to central belly pale pinkish-buff; larger black bib extends to central breast***. **Voice** Song a sweet, piping repetition of two or three notes with alternated stress, *tee-chu, tee-chu* ... or *chu-wee, chu-wee* ..., usually recalling Great Tit but more feeble; calls include a fluty, rather melancholy *puuht*, fine *pseet* and Goldcrest-like *si-si-si*.

Great Tit *Parus major* RB

L 14. Very common and widespread breeding resident. Breeds late Jan–May, often double-brooded to Jul; or later. **Habitat** Most common in coastal plains and foothills, also mountain forests to highest summits; wherever there are bushes and trees. Nest sites include tree holes, letterboxes, light fittings, pipes and tree stumps. **ID** *P. m. aphrodite* occurs. Note ***black head, white cheeks and broad, black stripe on centre of yellow underparts***. In juvenile, black areas sooty and cheeks yellowish, lacking dark lower border, and underparts dull yellow. **Voice** Highly variable, includes loud, metallic and clear *chink-chink* (recalling Common Chaffinch) or *tui-tui-tui*. Song distinctive, far-carrying and rather melodious, *tee-chu-tee-chu* ... or *tee-tee-chu, tee-tee-chu*. . ..

Eurasian Penduline Tit *Remiz pendulinus* PM, wv, FB?

L 11. Fairly common passage migrant and uncommon winter visitor. May breed occasionally, first reported 1983. Numbers vary. Autumn passage mainly late Oct–Nov, peaking early Nov; light spring passage mid Feb–mid Apr, with stragglers (or opportunist breeders) to early May. **Habitat** Tamarisks and live reedbeds, near fresh or brackish water. **ID** Long tail and ***finely pointed bill; chestnut mantle and wing-coverts***, blackish wings and tail feathers fringed white. Adult has ***black mask contrasting with pale grey crown and nape, and white throat***, female duller with smaller mask. Juvenile has cinnamon upperparts and lacks mask. **Voice** Commonest call a soft, rather quiet, ***drawn-out and falling, Robin-like, seeeeee***.

Bearded Reedling *Panurus biarmicus* pm, wv

L 17. Rare passage migrant and winter visitor, Oct–Mar. Formerly more common, has declined since mid 1980s. **Habitat** Reedbeds. **ID** ***Unmistakable, cinnamon-brown with very long tail***, short wings and weak, whirring flight, on rapid wingbeats, usually low over reeds in straggling disorderly parties. ***Male has blue-grey head, black moustachial and undertail-coverts***, both absent in brown-headed female and juvenile. Feeds quietly in reeds. Occasionally makes high-level aerial sorties in autumn and winter. **Voice** Characteristic pinging *tching*, also frequent *ptew* or *tlek*. [Alt: Bearded Tit.]

♂
♀
Hypocolius
ad
cypriotes
Coal Tit
ad
ater
ad
Great Tit
juv
♂
Eurasian Penduline Tit
♂
♀
Bearded Reedling

PLATE 62: COMMON LARKS

Calandra Lark *Melanocorypha calandra* RB, pm, wv

L 20. Locally common breeding resident, scarce passage migrant and occasional winter visitor. Passage or local movements late Feb–Apr and mid Sep–early Nov. Forms flocks in autumn. Breeds Dec–May. **Habitat** Ploughed fields and irrigated crops at low elevation. Breeds in extensive cereal fields. **ID** Large, heavy-billed lark with shortish tail. In flight *blackish underwing with conspicuous white trailing edge and white-sided tail*. On ground, *swollen yellowish-horn bill, black patches at sides of lower throat* and variable whitish supercilium. Smaller Bimaculated Lark has paler underwing without conspicuous white trailing edge, lacks white on sides of tail. Undulating flight is low with deliberate 'wader-like' wingbeats. **Voice** Flight call a harsh, rolling *terrelet*; also Eurasian Skylark-like note. Song, delivered in high circling flight on stiff deliberate wingbeats, richer and more powerful than Skylark, with much mimicry; can recall Corn Bunting.

Greater Short-toed Lark *Calandrella brachydactyla* PM, ob

L 14. Common passage migrant, occasional breeder. Spring passage late Feb–May, peaking mid Mar–mid Apr (often in large flocks); irregular late Aug–Nov, mainly Sep–early Oct. Breeds Jun–mid Aug; success partly depends on rainfall. **Habitat** Cultivated fields and gravel plains. **ID** Races *brachydactyla* and *hermonensis* occur, and probably other subspecies too. Colour variation and overlap in features evident among individuals. Small sandy lark with streaked upperparts, buffish-white *generally unstreaked underparts with variable black patch at side of neck* (adult); stoutish pointed bill, *long tertials almost covering wingtip*; bold median coverts. Streaked crown greyish, tinged rufous in some males. Upperparts sandy, grey-brown or rusty. From Lesser Short-toed Lark by variable *black neck patches, long tertials and longer bill*. Juvenile lacks neck patches, less streaky or has blotchy breast, much like Lesser Short-toed. Flight undulating; in dense, low-flying flocks. **Voice** Flight call a sparrow-like *tjirp*, and *drelit* similar to Common House Martin. Song unmusical, repetitive, in circling flight.

Crested Lark *Galerida cristata* RB

L 17. Common breeding resident. Breeds mid Feb–Jun. **Habitat** Coasts and foothills, low-intensity agriculture, marginal land, dusty field edges and salt flats. **ID** Stocky and short-tailed with *long spiky crest*. Upperparts sandy-grey, diffusely streaked darker on mantle; breast heavily streaked. Flight flapping, *broad wings rusty-buff below; tail blackish-brown with cinnamon sides. Lacks white trailing edge to wing*. **Voice** Clear *du-ee*, also variable fluty *ee* or *uu* sounds. Song sweet and plaintive with phrases repeated; slower and shorter than Eurasian Skylark, from exposed perch or high in air.

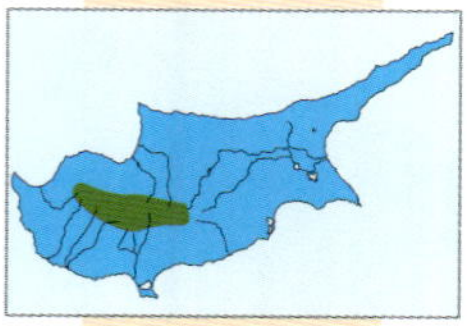

Woodlark *Lullula arborea* WV, pm, rb

L 15. Common winter visitor, fairly common passage migrant and breeding resident. Winter visitors (mainly from Central Europe) late Oct–mid Mar; passage late Jan–Feb and mid Nov–early Dec. Breeds Mar–Jul. **Habitat** Bare ground in olive and almond groves, vineyards, roadsides and recently graded fields. Breeds on open hillsides with scattered pines. **ID** Small, short-tailed lark, streaked on upperparts and crown with contrasting head markings and rusty ear-coverts bordered by *long, broad buffish-white supercilia* almost meeting on nape; diagnostic *black-and-white marking on leading forewing. Deep, undulating flight with noticeably broad wings and short tail*, shows white on tail corners and tip, wings lack white trailing edge. Often perches; on ground permits close approach. **Voice** Melodic song from perch or in flight of yodelling phrases is distinctive, *lee-lee-lee-lililililiIululu-luIooloo-eelu-eelu-eelu-eelu*. In flight, *characteristic yodel deedlui*.

Eurasian Skylark *Alauda arvensis* WV, PM, ob

L 18. Common winter visitor and passage migrant. Occurs Sep–Apr, mainly from Oct, peaking Nov–early Dec. Numbers decline Mar–Apr. Usually in medium-sized flocks. May breed occasionally, first reported 1997. **Habitat** Damp grassy meadows, cereal fields and ploughed areas; farmland on coasts and in foothills (to c.800m). **ID** Medium-sized with grey to sandy upperparts streaked dark, with warm brown edges to tertials and coverts in fresh plumage; heavily streaked on breast; can show small crest. *In flight, distinct whitish trailing edge to wings* and broad triangular *tail with white sides*. **Voice** Song, in fluttering flight high in air, a continuous stream of trills and warbles; when flushed and in flight, a variable rolled *chrriup*; or *trruwee*.

ad
Calandra Lark
ad
brachydactyla
ad
ad
hermonensis
Greater Short-toed Lark
Crested Lark
ad
Woodlark
ad
Eurasian Skylark

PLATE 63: RARE AND UNCOMMON LARKS

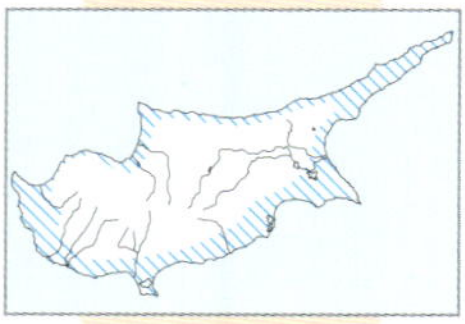

Bimaculated Lark *Melanocorypha bimaculata* pm

L 16. Uncommon passage migrant. Occurs Mar–early May, peaking late Mar–Apr. Infrequent Sep–Oct. **Habitat** Dry marginal land, recently ploughed fields and weedy coastal areas. Often with Greater Short-toed Lark. **ID** Resembles small, shorter-tailed Calandra Lark; in flight has ***dull grey-brown underwing without clear white trailing edge, and white-tipped tail*** with buff-brown outer feathers. On ground, ***upperparts show more prominent scaling*** and head pattern is more contrasting than Calandra, with rusty cheeks, pronounced ***long white supercilia and dark lores*** giving capped appearance. **Voice** Flight call recalls Calandra, *trrelit*, with rather gravelly scrunching; also a Greater Short-toed Lark-like *dre-lit*. Often retiring and overlooked, remaining still for long periods and appears wary.

Bar-tailed Lark *Ammomanes cincturus* v

L 13. Vagrant. Records Mar–Apr. **Habitat** Desert nomad, prefers sandy or gravel areas with sparse vegetation; mainly south-facing capes and headlands. **ID** Resembles Desert Lark but smaller, with more upright stance, rounder head, ***smaller bill, clear-cut black band to tip of reddish-brown tail***, and rufous-buff ***wings with blackish tips***. Race *arenicola* (Near East/Arabia) has unstreaked pale sandy-rufous upperparts. Runs fast with abrupt stops. From Dunn's Lark by ***much smaller bill, different tail pattern, unstreaked upperparts and crown, and blackish wingtips projecting beyond tertials***.

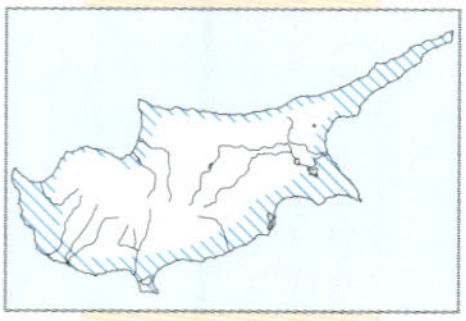

Lesser Short-toed Lark *Alaudala rufescens* pm, wv

L 13. Very scarce, irregular passage migrant and winter visitor. Mainly late Oct–early Apr, passage late Mar–early Apr and Sep–Nov. Movements rather complex and not well understood; several races reported, including *heinei* and *aharonii*. **Habitat** Salt flats, gravel plains and ploughed fields. **ID** Resembles Greater Short-toed Lark; slightly smaller but stockier, with ***stubbier bill, bulging 'jowl', more distinctly streaked breast without black patches at sides, and wingtips clearly exposed beyond tertials*** (beware Greater Short-toed with worn tertials). Upperparts sandy-grey to rufous-brown; crown occasionally tinged rufous; supercilium less noticeable than Greater Short-toed but pattern below eye bolder. Outer tail feathers white. Juvenile attains adult plumage from Aug. **Voice** Flight call abrupt, ***fast dry staccato prrrrt*** or *prrr-rrr-rrr*; also Eurasian Skylark-like *drrie* and quick *dreeup* (other short calls can be similar to Greater Short-toed).

(African) Dunn's Lark *Eremalauda dunni* v

L 14. Vagrant. Dunn's Lark formerly comprised two races – *dunni* and *eremodites* (Arabian Lark) – treated here as species. Ranges across North Africa from Sudan to Morocco. **ID** Small, sandy rufous-brown with ***dark streaks on crown and, vaguely, on mantle; black sides to sandy tail and pinkish bill with curved tip. Long tertials almost reach wingtip***. Broad whitish eye-ring bordered below by dark line, with dark moustachial and line behind eye. Paler, more orange-buff than Arabian Lark, face pattern plainer, with less prominent teardrop in front of eye, an indistinct supercilium and stubbier bill. Broad, rounded wings and short tail in flight. Appears large-headed, with upright stance; runs fast over short distances with sudden stops. **Voice** Flight call drawn-out, soft *wazz* or *ziup*; also thin liquid *prrrp*; alarm a rising *chee-opp*, *tu-wep* or *chup-chup-chup*.

Horned Lark *Eremophila alpestris* v

L 16. Vagrant. **Habitat** Most likely on headlands, although has occurred in Troodos Forest. **ID** Confusable with smaller Temminck's Lark, but has pinker nape; male has white or ***yellowish and black face markings; black cheeks connecting black breast-band***. Female lacks black 'horns' and has washed-out head markings. Upperparts greyish, almost unstreaked in Middle East races; tertials dark-centred, offering little contrast with wingtip (unlike Temminck's). In flight wings uniform, black tail has pinkish-brown central feathers and white sides. [Alt: Shorelark.]

Temminck's Lark *Eremophila bilopha* v

L 14. Vagrant. **ID** Resembles Horned Lark but smaller with more contrasting, paler sandy upperparts including tertials, which contrast with black wingtips; diffusely rufous lower hindneck; facial markings always white, and black cheeks and breast-band never join. Juvenile lacks black on head, throat and breast, is uniform sandy-cinnamon above and easily mistaken for other larks. **Voice** Squeaky jingling similar to Horned Lark.

Bimaculated Lark

Bar-tailed Lark

Lesser Short-toed Lark

(African) Dunn's Lark

Horned Lark

Temminck's Lark

PLATE 64: SWALLOWS AND MARTINS

Sand Martin *Riparia riparia* PM

L 12. Very common autumn passage migrant, less common in spring. Irregular late Feb–mid Jun (rather scarce until Apr), peaking mid Apr–early May; common mid Aug–early Nov, peaking Sep–early Oct. Occasional in summer and winter. **Habitat** Widespread on migration, over pools, dams, reedbeds and irrigated fields. **ID** Small, dull *brown-and-white martin*; white underparts with well-marked *brown breast-band* (paler and indistinct in juveniles) and longish, distinctly forked tail. Juvenile scaly, with buff upperparts and ill-defined breast-band. **Voice** Flight call a vowel-less, rasping repeated *tschr* or *zrrrr-zrrrr*, persistent and slightly buzzy in groups; louder *chirr* in alarm.

Barn Swallow *Hirundo rustica* PM, MB

L 16 (including tail streamers). Very common passage migrant and breeding visitor. Spring late Jan–early May, peaking mid Mar–late Apr; autumn late Jun–Nov, mainly late Aug–Oct, occasional in winter. Breeds from Feb, juveniles from mid Apr. Often double brooded to July. **Habitat** Widespread, congregating near coasts and wetlands, and roosts in reedbeds. Nests inside buildings, porches, car ports and verandas, or anywhere with suitable ledges. **ID** *Bluish-black upperparts, long tail streamers* (small *white patches in tail* when spread), *chestnut forehead and throat, and solid dark breast-band*; underparts buffish-white including underwing-coverts. Some races and variants have underparts reddish-buff. Juvenile lacks tail streamers, has brownish breast-band and pale forehead. Flight strong and elegant with much banking, often foraging low. Perches on wires with other hirundines. **Voice** A melodious twittering; contact call *witt-witt*, alarm a disyllabic *tsi-wit*.

Eurasian Crag Martin *Ptyonoprogne rupestris* rb, pm, wv

L 14.5. Uncommon breeding resident, passage migrant and winter visitor. Spring mostly Feb–early May, especially Apr; autumn Sep–early Nov, peaking late Oct. Variable numbers in winter, sometimes large groups, Nov–mid Mar. Possible weather-related altitudinal movements. Breeds Apr–mid Jul. **Habitat** Steep slopes and ravines in montane forests; feeds along streams and watercourses. Passage over capes and headlands; in winter dams and reservoirs, sheltered cliffs, south-facing castle walls and valleys on low ground. Nests in caves and beneath overhangs, eaves and bridges. **ID** Large broad-winged, brown martin, almost square-tailed; *heavily built, with dark upperparts lacking greyish tinge to rump; cheeks fairly dark, usually contrasting with pale throat and underparts; fine spots on chin and throat* (hard to see). Underparts buffish-grey, darker towards undertail; *underwing-coverts distinctly darker than flight feathers*. Flight more swooping and diving than other hirundines. **Voice** Song a quiet twittering; in flight, a short *chip* or *chirr*.

Common House Martin *Delichon urbicum* MB, PM

L 12.5. Very common breeding visitor and passage migrant. Spring from late Jan, peaking mid Mar–mid May. Breeds Feb–May, or earlier. Large flocks feed over forests Jun–Aug. Autumn late Jul–Oct, peaking Aug–early Sep. **Habitat** Mountains, hills and villages. Nests on sea cliffs and beneath eaves of buildings at all altitudes, often in large colonies. **ID** Smaller than Barn Swallow; *bluish-black upperparts with striking white rump and all-white underparts*; short, forked tail with white undertail-coverts and rather dark underwing. Juvenile duller with brownish wash on head and breast-sides. Flight more fluttering than Barn Swallow with long glides often high in air. Frequently perches on wires. Gregarious, migrating in flocks. **Voice** Commonest flight call a short dry warbling *prrlit*, sounding conversational. Repeated high trilled *scheeer* in alarm.

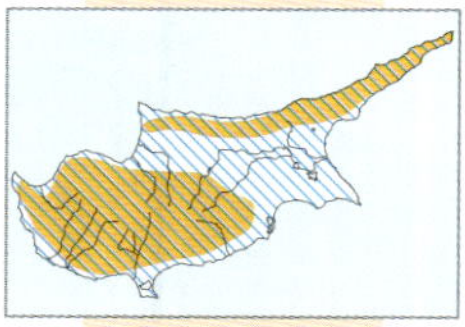

Red-rumped Swallow *Cecropis daurica* PM, MB

L 17 (including tail streamers). Common passage migrant and breeding visitor. Common in west of the island. Occurs mid Feb–early Dec, mainly Mar–early May, peaking mid Apr; fewer Aug–early Nov. Breeds mid Mar–Jun. **Habitat** Fertile valleys, hillsides and rocky foothills; headlands on passage. Nests in culverts and pipes beneath roads and tracks, and under bridges. **ID** Resembles Barn Swallow but has *rufous collar and broad pale rufous rump*; underparts buffish-white, faintly streaked (variable), *undertail-coverts strikingly black; lacks breast-band and white in tail*. Juvenile browner with paler collar and rump, and shorter tail. Flight slow and graceful, frequently gliding for longer periods than Barn Swallow, with tail flared or closed. **Voice** Quieter shorter song than Barn Swallow; in flight, a short soft nasal *tweit*, reminiscent of House Sparrow.

Sand Martin
ad ♂
Barn Swallow
juv
Eurasian Crag Martin
Common House Martin
ad
Red-rumped Swallow

Cetti's Warbler *Cettia cetti* RB

L 13.5. Abundant and widespread breeding resident. Has increased in range since 1970s. Sings from Oct, breeds late Dec–Jul. **Habitat** Low dense vegetation, reedbeds and overgrown areas by rivers and ditches at most altitudes; also bushy salt scrub and drier, sheltered locations. **ID** Race *orientalis* occurs. Short-winged, ***round-tailed***, robust-looking, ***warm brown warbler***, keeps well-concealed in thickets and more often heard than seen; ***song distinctive***. Off-white below, tinged rufous on flanks and breast-sides; ***vent and undertail-coverts brown with white-fringed feathers***; uniform chestnut-brown tail often jerked or cocked; thin ***distinct white supercilium***. **Voice** Abruptly delivered song from well-concealed perch a ***sudden, loud, explosive outburst plit – plitiplitipliti –plhi-(pliti)***. Alarm call a hard, forceful *tlitt*, or rapid, rattled Eurasian Wren-like *tlitt-tlitt-tlitt....*

Zitting Cisticola *Cisticola juncidis* RB

L 10. Common and widespread breeding resident. Song and display from late Dec, breeds mid Jan–early Jun, sometimes to Aug. **Habitat** Meadows, cereal fields, roadside verges, marshy edges and low-intensity farmland, from coastal plains to treeline. Purse-shaped nest in bush or long grass. **ID** Small, short-tailed warbler, often seen in song flight. Crown and upperparts clearly streaked yellow-buff and black; face pale with prominent dark eye; tail rounded, often fanned in flight, with black-and-white tips, most obvious from below in song flight. **Voice** Song diagnostic, in bounding circular flight; consists of *tzip* notes repeated at regular intervals of 0.5–1.0 second, with one *tzip* per bound; also a short *kwit* from cover. [Alt: Fan-tailed Warbler.]

Common Grasshopper Warbler *Locustella naevia* V

L 12.5. Vagrant. **ID** Small, rather dark, skulking warbler with obviously rounded tail when flushed, often from underfoot! Seeks cover in short, low, jerky flight. ***Upperparts olive-brown, heavily streaked*** (but appearing obscure in flight) with ***faint supercilium***; underparts dirty white or yellowish with streaked rear flanks and undertail-coverts, diffusely so on darker breast. Smaller Zitting Cisticola is streaked black and yellowish with short, well-patterned tail. Cetti's and Savi's Warblers have plain upperparts. **Voice** Very occasionally sings on migration. Song a dry fishing rod-like reeling, longer, more sibilant and less whirring than Savi's Warbler; infrequently gives a short *chik* call.

River Warbler *Locustella fluviatilis* V

L 13. Vagrant. **ID** Secretive grey-brown warbler, with ***unstreaked upperparts, diffusely streaked breast and long undertail-coverts with brownish feathers tipped white***. Confusable with Savi's Warbler but darker and greyer, being colder, darker olive-brown above, with much more obvious darker spots on breast (diffuse in Savi's); generally keeps low in vegetation, and will walk on ground with slightly cocked tail. **Voice** *Characteristic song*, occasionally heard on spring migration: a mechanical 'steam-train' rhythm, *tze tze tze tze tze tze....*

Savi's Warbler *Locustella luscinioides* pm

L 14. Uncommon passage migrant. Mainly spring mid Mar–Apr, peaking mid Apr; rare or overlooked Aug–Oct. Also other months. **Habitat** Reedbeds, damp overgrown areas and meadows. **ID** Nominate *luscinioides* usually occurs. A large *Locustella*, slightly larger than Eurasian Reed Warbler. Plumage warm brown and ***breast only lightly marked with very diffuse faint brownish lines of spots***; long undertail-coverts sometimes tipped pale (certainly so in eastern race *fusca* which may also occur) but less distinct than in River Warbler. Often on ground, walking slowly and stealthily with horizontal stance and jerky movements, graduated tail raised; also often bobbed (a character of all *Locustella*). **Voice** Song, uttered from reed stem or bush, by day or night, similar to Common Grasshopper Warbler, a monotonous, soft reeling, ***lower in pitch and faster.*** Call also distinctive, a loud, hard, twinky *pitch* repeated two or three times. Migrants often sing in spring.

ad
ad
Cetti's Warbler
Zitting Cisticola
Common Grasshopper Warbler
ad
ad
River Warbler
ad
luscinioides
fusca
Savi's Warbler

PLATE 66: *PHYLLOSCOPUS* WARBLERS I

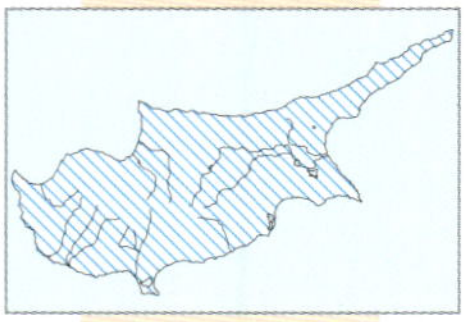

Willow Warbler *Phylloscopus trochilus* PM

L 11. Very common autumn passage migrant, less so in spring. Passage mid Mar–May, peaking in Apr; autumn common mid Aug–mid Nov (when yellow-toned juveniles predominate), peaking Sep–mid Oct. **Habitat** Weedy meadows and roadsides; bushes and scrub. **ID** Similar to Common Chiffchaff, with *longer primary projection; darker, pale-edged flight feathers and tertials; yellow wash on throat and breast; more distinct supercilium and pale-centred ear-coverts*. Race *acredula* often appears olive and yellow, commonest in spring. *Legs pale pinkish-brown*; tail-dipping much less pronounced. *First-winter has distinctly yellow underparts*. **Voice** Song a pleasant *descending whistled verse that dies away at end*. Call a rising, disyllabic *hoo-eet*.

Common Chiffchaff *Phylloscopus collybita* PM, WV

L 11. Very common passage migrant and winter visitor. Occurs late Sep–Apr, mainly late Oct–Mar. **Habitat** Trees, bushes, reedbeds and crops. Three races predominate (with intermediates that are indistinguishable in field): *collybita*, *brevirostris* and *abietinus*. All show *short supercilia*, broken eye-ring (conspicuous in autumn) and *dark legs; tail frequently dipped*.

P. c. collybita (not illustrated) is a fairly common passage migrant and winter visitor. **ID** Brownish-olive above, more olive on rump; short, weak supercilium. Underparts sullied white, streaked yellow on breast, buff suffusion on flanks; undertail-coverts yellowish-white; underwing-coverts and wing bend yellow. Wings and tail have indistinct yellowish-olive fringes. Weak bill dark horn, yellowish-brown lower mandible base. **Voice** Calls plaintive *hoo-ee* or *huit*, with rising inflection, more monosyllabic than Willow Warbler; also downward-inflected disyllabic *sweeoo*.

P. c. brevirostris (breeding range includes north Turkey) is common mid Nov–Feb. **ID** *Brown-tinged above (especially crown) and on breast-sides*, with more distinct supercilium (especially above and in front of eye), *off-white undertail-coverts* and shorter primary projection. **Voice** Plaintive call, *peep* or *swee*, similar to *tristis* (see below).

P. c. abietinus 'Scandinavian' Chiffchaff is a very common passage migrant and winter visitor. Occurs Oct–mid Nov and Mar–Apr (including *abietinus/collybita* intergrades). **ID** Some paler and greyer above than *collybita*; buff and yellow on breast reduced, undertail-coverts whiter. **Voice** As *collybita*.

P. c. tristis 'Siberian' Chiffchaff (sometimes given specific status) is a rare passage migrant or winter visitor. Mostly Nov–early Feb. **ID** *Greyish-brown above, lacking yellow except on wing bend and underwing, olive-green tones on rump, wing-coverts, scapulars and edges of flight and tail feathers. Notable thin supercilium, off-white to buff, rarely with trace of yellow*, ear-coverts with rusty hue; underparts off-white, suffused buffy breast-sides and flanks; *legs and bill almost black*. In fresh plumage *may show indistinct wingbar* on greater coverts. **Voice** Call a monosyllabic, plaintive *eep*, recalling Dunnock.

Mountain Chiffchaff *Phylloscopus sindianus* V

L 11. Vagrant. Race *lorenzii* has occurred. **ID** Recalls *tristis* Common Chiffchaff, *lacking obvious yellow or green* (flight feathers have subdued olive fringes, also faint tinge on scapulars; axillaries and underwing-coverts pale yellow). *Upperparts earth-brown*, brown crown darker, with *broad white supercilium, broader in front of eye*, buff behind it; white lower eyelid, *dark lores* and buff ear-coverts. *Underparts off-white*; white bib surrounded by pinkish-buff, *breast-sides and flanks tinged rusty*. Lower mandible base yellow; legs and strong feet black. Appears short-winged; long and broad square tail. Tail-dipping and behaviour as Common Chiffchaff. **Voice** Plaintive call, *biii* or *swee*.

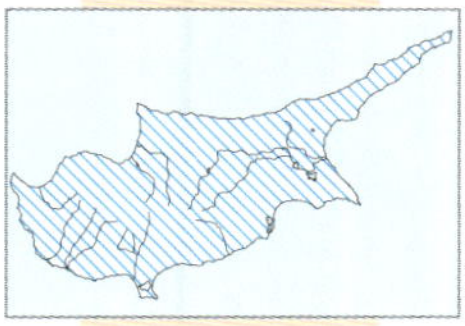

Eastern Bonelli's Warbler *Phylloscopus orientalis* PM

L 12. Fairly common spring passage migrant, rare in autumn. Occurs Mar–mid May and infrequently Jul–mid Oct. **Habitat** Orchards, mixed woodland and conifers. **ID** Size as Willow Warbler but *mantle and more rounded head distinctly grey*, contrasting dark tertials fringed silvery-grey, with *obvious green edges to wing and tail feathers*; underparts silky-white; *face 'washed-out' with large dark eye* accentuated by narrow, *complete white eye-ring, pale lores and ear-coverts, and indistinct supercilium*. If seen well, *yellowish-green rump distinctive*. Confusable with pale Chiffchaff, but note plain head and longer primary projection. **Voice** Often located by repeated call: a short, flat, *Crossbill-like 'chip'*.

nominate
1st-winter
acredula
Willow Warbler
tristis
brevirostris
abietinus
Common Chiffchaff
ad
ad
1st-winter
Eastern Bonelli's Warbler
Mountain Chiffchaff

Wood Warbler *Phylloscopus sibilatrix* PM

L 12.5. Fairly common passage migrant. Occurs late Mar–mid May, peaking Apr–early May; infrequent late Aug–early Nov. **Habitat** Large mature trees, well-vegetated river valleys, orchards and pine forests; regular in mountain juniper and pines in autumn. **ID** Large, ***long-winged***, brightly coloured *Phylloscopus*: ***vivid green upperparts with yellow supercilium, ear-coverts, throat and upper breast, contrasting with gleaming white underparts; yellow-fringed secondaries form pale panel contrasting with whitish-fringed blackish tertials***; black alula, yellow bill and pale legs. From Green Warbler by different wing pattern, including lack of wingbar; from other *Phylloscopus* also by very long wings, striking plumage, particularly head pattern and brightly fringed secondaries. Rare greyer individuals, recalling Eastern Bonelli's Warbler, by strength of supercilium, long wings (still contrastingly fringed) and larger size. Rather active, often high in trees. **Voice** Usually silent on migration.

Yellow-browed Warbler *Phylloscopus inornatus* V

L 10. Vagrant. Records late Feb–early Apr and late Sep–Nov, peaking late Oct. **Habitat** Trees and low bushes. **ID** A small warbler with very long, pronounced ***yellow-white supercilium*** and ***two distinct yellow-white wingbars*** that contrast with dark-centred greater coverts. Otherwise, bright olive-green above and clean whitish below. Note whitish edges to tertials; occasionally shows faint paler crown-stripe. In worn plumage greyer with paler wingbars, and tertial fringes narrower and whiter, thus similar to Hume's Leaf Warbler, which see for separation. **Voice** Song (may be heard on migration/winter) high-pitched *tsee tseoo-tseee*, call high-pitched lisping *tsweest* or *weest* with distinct upward inflection, frequently an almost disyllabic *wii-ist*.

Hume's Leaf Warbler *Phylloscopus humei* No records

L 10. A possible accidental visitor. **Habitat** Trees and bushes. **ID** Confusable with Yellow-browed Warbler; ***generally greyer, duller, less yellow-green*** with different voice. In fresh plumage, upperparts and crown greyish-olive with very long, pronounced ***buffish-white supercilium***, often broadest above and in front of eye; two wingbars, lower one distinct, ***upper short and ill-defined or absent*** (both distinct in Yellow-browed); ***underparts dusky-white, tinged buffish on ear-coverts, lower throat and neck-sides; bill dark*** (but pale base from below). Occasionally shows faint paler crown-stripe. In winter/spring, worn plumage greyer, wingbars and tertial edges narrower, with upper wingbar often lacking. **Voice** Calls, sweet *wesoo* or shorter *dweed*, both ***lacking upward inflection***; in winter more disyllabic, clean *sooit*, recalling compressed Spotted Redshank. [Alt: Hume's Yellow-browed Warbler.]

Green Warbler *Phylloscopus nitidus* V

L 11. Vagrant. **Habitat** Trees and bushes. **ID** A bright olive-green and yellow warbler, similar in size to Common Chiffchaff, with broad ***sulphur-yellow supercilium from base of pale bill to behind ear-coverts***, short narrow wingbar on greater coverts and ***sulphur-yellow throat and upper breast***. From similar Greenish Warbler *P. trochiloides* (not recorded in Cyprus) by greener upperparts, sulphur-yellow throat, upper breast, supercilium and ear-coverts contrasting with rather dark eyestripe. In worn plumage Green Warbler can be more greyish, almost lacking yellow tones, and wingbar may be lost; separation from Greenish Warbler is then very difficult, but from most other *Phylloscopus* by striking head pattern, pale bill and voice. **Voice** Call a White Wagtail-like *che-wee*.

Dusky Warbler *Phylloscopus fuscatus* V

L 11. Vagrant. **Habitat** Low cover, or scrub, often in moist or marshy localities. **ID** Small, ***dark brown*** *Phylloscopus* with distinct head pattern resembling brownish Common Chiffchaff. ***Upperparts plain brown with rather distinct long, pale rufous-brown or whitish supercilium***, narrow and often palest in front of eye. Underparts off-white, ***tinged pale rufous-buff on breast, flanks and undertail-coverts. Fine bill with largely yellowish lower mandible; thin legs flesh-brown***. From *tristis* Common Chiffchaff by call, paler legs, paler bill (blackish in *tristis*), often rufous-tinged vent and undertail-coverts (not pale grey/off-white as Common Chiffchaff), also long striking supercilium (duller and shorter in *tristis*). Often on ground. **Voice** Loud, hard, abrupt *chek* or *tack*, recalling Lesser Whitethroat, often persistently given.

ad
Wood Warbler
ad fresh
Yellow-browed Warbler
ad fresh
1st-winter
1st-winter
Hume's Leaf Warbler
ad fresh
1st-winter
Green Warbler
Dusky Warbler
ad

Basra Reed Warbler *Acrocephalus griseldis* V

L 15. Vagrant. **Habitat** Reedbeds; on migration also bushes and other cover. **ID** Smaller than Great Reed, but larger than Eurasian Reed Warbler. From latter also by ***longer bill***, better-marked supercilium, less warm-coloured upperparts and ***darker legs***. From larger Great Reed by ***narrower bill***, short but ***more conspicuous supercilium (especially above and behind eye)***, greyer, less warm upperparts, ***whiter underparts (especially flanks and undertail)***, shorter, more square-cut tail and shorter, ***greyer legs***. **Voice** Call note a harsh *chaarr*, louder than Eurasian Reed.

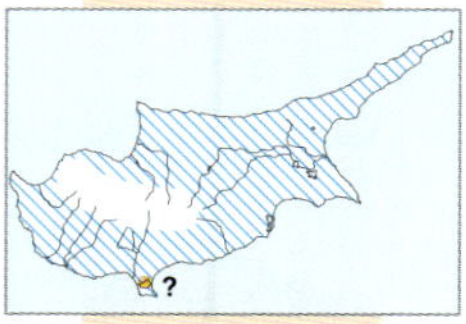

Great Reed Warbler *Acrocephalus arundinaceus* PM, ob

L 19. Common spring passage migrant, less common in autumn; occasional breeder. Numbers vary. Occurs Mar–May, peaking mid–late Apr, with stragglers to early Jun some years. Autumn passage less obvious, mainly mid Aug–Oct, occasionally later. Migrants often sing in spring, but breeding is rare and seldom proven. **Habitat** Reedbeds, also overgrown damp vegetation. **ID** ***Very large, much larger than Eurasian Reed Warbler with more prominent supercilium and noticeably heavier bill***. Contrasting buffier rump in flight. Size makes it easy to locate when moving ponderously through vegetation. Compared to Clamorous Reed Warbler *A. stentoreus* (a potential vagrant, illustrated) Great Reed has ***longer primary projection with 7–8 primary tips visible***, proportionately shorter tail and shorter, slightly stouter bill. **Voice** Loud and powerful song with repetitive character; common phrase *trr-trr-, karra-karra-karra, kreee-kreee-kreee*.

Moustached Warbler *Acrocephalus melanopogon* WV

L 13. Uncommon winter visitor. Rather local. Reported late Sep–Apr, mainly late Oct–Feb, peaking early Nov. Much confusion with Sedge Warbler in spring and autumn, when migration periods overlap. **Habitat** Healthy reedbeds and overgrown marshy edges. **ID** Nominate *melanopogon* predominates. Similar to Sedge Warbler but has broad, ***clear-cut white supercilium, ending squarely on side of nape and separating blackish crown and greyish ear-coverts*** (Sedge has buffier supercilium, paler central crown, especially in fresh plumage, and paler ear-coverts). Also ***more uniform, rufous-brown upperparts*** (less obviously streaked than Sedge, especially on coverts), with ***whiter throat***, warmer brown wash on flanks and, sometimes, finely streaked warm breast-band. Subspecies *mimica* (Turkey to Iran) is slightly less rufous above with more streaked crown but still ***darker above than Sedge Warbler, with white supercilium and distinctly shorter primary projection***. Keeps low in vegetation or on ground, where unobtrusive, ***often cocking and flicking tail nervously***, also when singing. **Voice** Migrants often sing before departure. Song like a sweet Eurasian Reed Warbler (but more variable) with intermittent ***characteristic Common Nightingale-like lu lu lu***. Call a loud *trr-trr*, soft, short *tcht*, longer *trr-trrrrr* and a hard *tack* (like tapping two stones together).

Sedge Warbler *Acrocephalus schoenobaenus* PM

L 13. Common spring passage migrant, less numerous in autumn. Spring passage Mar–May, peaking in Apr; size of autumn passage dependant on available wetlands, late Jul–Nov, peaking late Sep–Oct. **Habitat** Bushy riversides, irrigated fields, ditches and wetlands with reedbeds and grasses; usually at lower elevations. **ID** Buffish, streaked, wetland-dwelling warbler with buffish-white supercilium, confusable with Moustached and Aquatic Warblers. ***From former by buffier, less square-cut supercilium, slightly paler crown, paler ear-coverts and buffier, more streaked, upperparts, which merge into warmer-coloured rump, and distinctly longer primary projection***. From Aquatic by less striped head and much less well-marked upperparts. First-winter Sedge Warbler is yellower than adult, shows paler centre to crown and fine spotting on breast. **Voice** Quite vocal on spring migration, calling and singing. Song similar to Eurasian Reed Warbler but faster and more erratic, often rising and falling in pitch; frequently mimics and often starts song with rapid *trr* notes; often song flights. Call a hard *chek* and fast, churring *trrr*.

ad
Basra Reed Warbler
ad
Great Reed Warbler
ad
Clamorous Reed Warbler
ad
melanopogon
mimica
ad
Moustached Warbler
1st-winter
ad
Sedge Warbler

Paddyfield Warbler *Acrocephalus agricola* V

L 12. Vagrant. **Habitat** Reedbeds and scrub; also damp fields on passage. **ID** Similar to, but slightly smaller than, Eurasian Reed Warbler and differs in ***shorter bill, more prominent whitish-buff supercilium from bill to well behind eye*** (where most conspicuous), ***often with darkish border above***, greyish shawl, dark-centred tertials with paler edges (unlike Blyth's Reed); ***shorter wings and longer tail, which when landing or on ground is often raised slightly and constantly flicked***. Underparts show warmer wash to flanks and undertail than other unstreaked *Acrocephalus*. Bill brown with flesh-coloured lower mandible but distal part often darker in adult (can appear dark-tipped). Crown feathers often raised; iris pale in adult. **Voice** Calls include a simple *chik, chik*, a tit-like *dwzzz* and a rolling *churrr*.

Blyth's Reed Warbler *Acrocephalus dumetorum* V

L 12.5. Vagrant. **Habitat** Dense bushy vegetation, mostly near water. **ID** ***Similar to Eurasian Reed and Marsh Warblers, and difficult to identify*** without experience or close prolonged views or photographs. Identified by combination of ***shorter, more rounded wings and short primary projection with six visible primary tips*** (7–8 in Eurasian Reed and Marsh Warblers), ***uniform upperparts with little or no rump contrast, plainer tertials***, near-concolorous alula, short supercilium, which often bulges in front of eye, and dark grey bill with flesh-coloured base to lower mandible. Sometimes flicks and fans tail (unlike Eurasian Reed and Marsh Warblers). Carriage usually more horizontal with head and bill inclined upwards in 'banana' posture. **Voice** Contact call, often given, a soft *thik* or *chck*, recalling Lesser Whitethroat.

Eurasian Reed Warbler *Acrocephalus scirpaceus* MB, PM

L 13. Common breeding visitor and passage migrant. *A. s. fuscus* (Caspian Reed Warbler) is a common breeding visitor and most migrants are believed to be this race. *A. s. scirpaceus* breeds in westernmost Turkey and is a passage migrant through Cyprus in unknown numbers. Occurs late Feb–Nov. Spring passage late Mar–early May, peaking early–mid Apr; autumn migration mid Aug–Oct very light and overlaps with post-breeding dispersal. Breeders increasing in range and numbers; nests Mar–May. **Habitat** Almost exclusively reedbeds; occasionally nests in long grass and tamarisk bushes. **ID** Reed-dwelling warbler with flat forehead and rather long bill; brown above with slight olive tinge to mantle and, especially, ***warmer rufous-brown rump***; buffish-white below with warm wash to flanks; short, rather indistinct supercilium. Similar to rarer Marsh Warbler, with similarly long wing projection but less obvious pale edges to tertials, ***longer bill and darker legs***. Caspian Reed Warbler greyer, less rufous above and whiter below. **Voice** Monotonous and fairly even-pitched mix of scratchy, grating and churring notes; ***resembles song of Sedge Warbler but slower and lacks changes in pitch and tempo***. Calls short *tchk* or hard, rolling *chrrrur*.

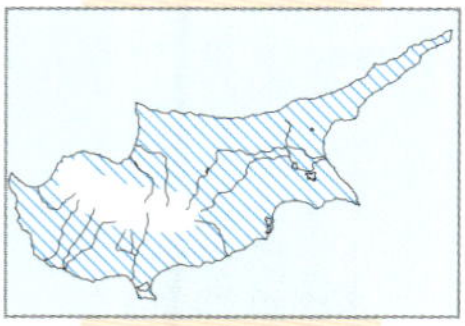

Marsh Warbler *Acrocephalus palustris* pm

L 13. Scarce to rare passage migrant. Very scarce Apr–mid May; less so Aug–Oct, mainly Aug. Probably overlooked due to its similarity to Eurasian Reed Warbler. **Habitat** Dense woody vegetation in ditches, thickets and streambanks; almost any cover on passage. **ID** Often difficult to identify. Brown above with ***slight olive tinge*** and marginally warmer rump; cream-buff below with slightly buffier flanks; short buffish white supercilium and pale eye-ring. Like all *Acrocephalus* has a rounded tail, long undertail-coverts and rather sloping forehead. ***Marsh Warbler best identified in spring by pale-fringed dark tertials and long primary projection, eight visible primaries, each with pale-fringed tip***. Also buffier, less warm, plumage than Eurasian Reed (although juvenile Marsh Warbler rustier and thus very similar), with paler legs and shorter bill, but latter hard to judge. **Voice** Markedly different song from Eurasian Reed Warbler; ***loud, lively, musical and full of mimicry***, mostly at very fast tempo, often including nasal *be-zec, be-zec*; when singing red gape often conspicuous. Calls short *chek* and distinctive short buzzy *terrrr*.

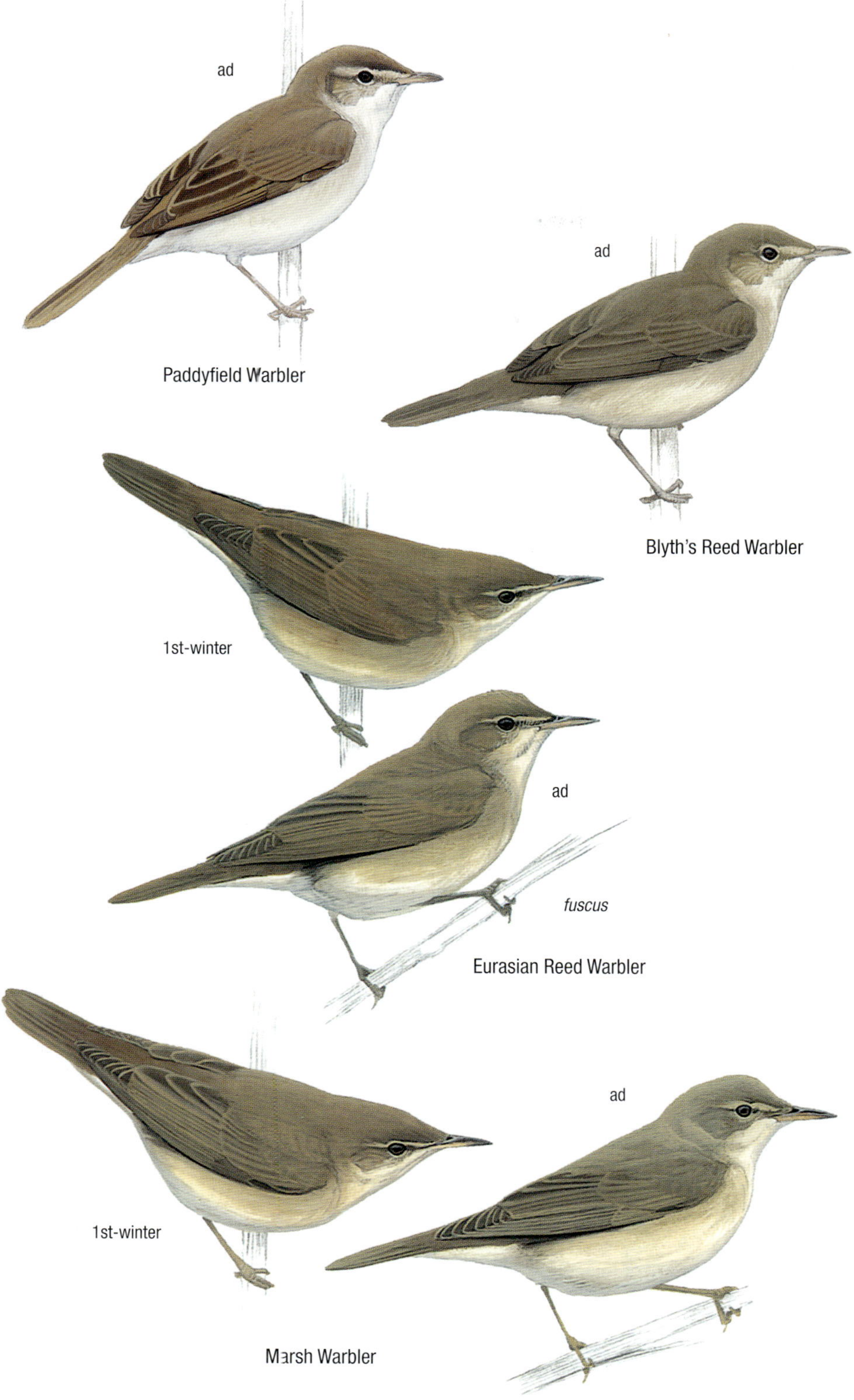
ad
Paddyfield Warbler
ad
Blyth's Reed Warbler
1st-winter
ad
fuscus
Eurasian Reed Warbler
1st-winter
ad
Marsh Warbler

Eastern Olivaceous Warbler *Iduna pallida* MB, PM

L 12.5. Very common breeding visitor and passage migrant. Occurs mid Mar–mid Oct. Spring passage, a few from mid Mar–mid Apr, peaking late Apr–May; light autumn passage mid Jul–Oct (with local breeders dispersing first). Breeds late Mar–Jul. **Habitat** Favours riverside bushes including tamarisk and brambles; olive and almond groves in sheltered valleys; breeds in deciduous forest valleys at all altitudes. **ID** Recalls *Acrocephalus* warblers, but greyer with square-ended or slightly rounded tail, and ***shorter undertail-coverts. Upperparts olive-brown, tinged greyish*** on head and mantle, underparts buffish-white (juvenile suffused yellowish at least on breast) with white throat; ***marked pale wing panel in spring at least***, tail brown inconspicuously fringed and tipped white on outer feathers. Pale eye-ring and variably distinct ***supercilium from bill to rear of eye; long, rather heavy, broad-based bill with yellow-pink lower mandible; legs greyish-brown***. Often deliberately flicks or ***pumps tail downwards*** (lacks circular tail movements of Upcher's Warbler). Usually feeds in canopy. Prefers to stay deep in cover even when singing. **Voice** Song like Eurasian Reed Warbler in quality and rhythm, repetitive and unmusical, but jaunty. Calls short *tch* or *tchek*, or more drawn-out *che-ch-ch* or agitated *trrrrr*.

Upcher's Warbler *Hippolais languida* v

L 14. Vagrant. **ID** Similar to Eastern Olivaceous Warbler but larger with stronger bill, and proportionately ***longer, much darker sooty-brown tail with broader white tips and edges, especially visible from below. Tail uniquely waved up and down, and sideways in circular movements***. Upperparts greyer in fresh plumage (more like Eastern Olivaceous in abraded and juvenile plumage) with darker wings and prominent pale panel on closed wings in fresh plumage. Head more rounded than Eastern Olivaceous, pale supercilium weaker and often extends behind eye (depending on lighting); legs cold grey. Larger Olive-tree Warbler has longer wing projection, powerful yellow-orange bill and short supercilium usually only visible in front of eye. Upcher's Warbler usually prefers mid and lower levels of bushes. **Voice** Call similar to Eastern Olivaceous, but mostly silent on passage.

Olive-tree Warbler *Hippolais olivetorum* pm

L 16. Rare passage migrant. Not annual and mostly in spring Apr–May, peaking late Apr; rarer in autumn late Aug–mid Sep. **Habitat** Scrub and bushes on passage. **ID** Large, brownish-grey with ***prominent wing panel and long, dagger-like, yellow-orange bill***; forehead flat, sometimes raises crown feathers, like Icterine Warbler. Head pattern weak, faint supercilium ***usually only in front of eye***. Underparts whitish-buff; off-white throat contrasting with rather dark head and ear-coverts. Undertail-coverts show darker 'V's, hard to see. Tail dark, edged and tipped whitish on outermost feathers; legs bluish-grey. Resembles Upcher's Warbler but larger, with stronger and paler bill, longer wing projection and more prominent wing panel. Often dips tail, sometimes waved and spread. **Voice** Call, often repeated, *tuc-tuc*.

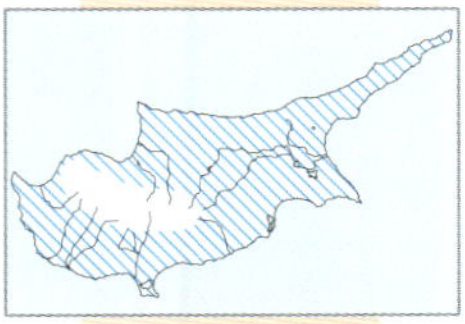

Icterine Warbler *Hippolais icterina* pm

L 13.5. Uncommon passage migrant. Numbers vary annually. Spring passage mid Apr–mid May; autumn early Aug–early Nov. **Habitat** Wooded and bushy areas. **ID** Fairly large yellow and olive warbler, with peaked crown, heavy yellow-orange bill (red gape when singing), ***very long wings with pale panel in fresh plumage (less so in autumn adult) and bluish-grey legs***. Bare-faced expression with conspicuous yellowish eye-ring and pale lores making eye rather prominent; yellowish supercilium short, rarely to rear ear-coverts. From *Phylloscopus* warblers by larger size, peaked crown, stronger bill, prominent wing panel, and lack of contrasting supercilium and eyestripe. Juvenile has browner upperparts and paler, more yellowish-white underparts with wing panel more conspicuous than worn adult. Occasionally has greyish-olive upperparts and whitish underparts, recalling Eastern Olivaceous or Upcher's Warblers, but note longer wing projection and voice. **Voice** Varied song is loud, long, with fast tempo, highly mimetic, sequences repeated two to five times, constantly interrupted by harsh chatter and nasal creaky 'violin' sounds, *de-de-dwiie* or *djehk-hyyii*; also includes characteristic trisyllabic *de-te-roy* (recalls Eurasian Siskin); alarm call, often given by migrants, a hoarse, sparrow-like *tettettettett*.

ad fresh
ad worn
Eastern Olivaceous Warbler
ad worn
ad fresh
Upcher's Warbler
ad worn
ad fresh
Olive-tree Warbler
1st-winter
ad
Icterine Warbler

Eurasian Blackcap *Sylvia atricapilla* PM, WV, ob?

L 14. Very common passage migrant and winter visitor. Spring passage significant, Mar–early May (or later), peaking Apr; autumn mid Aug–Nov, peaking late Oct. Winter visitors present Nov–Mar, when rather shy. Main target of poachers, especially east of Larnaca. Reportedly breeds, but poorly documented. **Habitat** Favours citrus, almond and olive groves at lower elevations, occasionally in montane forests. On passage anywhere with trees and bushes. **ID** Similar in size to Common Whitethroat; plain grey-brown above with distinctive solid ***black cap*** in adult male and ***warm brown cap*** in female and juvenile, lacks any white in outer tail feathers; these features prevent confusion with other warblers. Sings from cover. **Voice** Joyful, liquid song, rich and varied, ending in a few clear melodious fluted notes. Call a hard insistent *tack*, or repetition, similar to Lesser Whitethroat; also short *churr churr*.

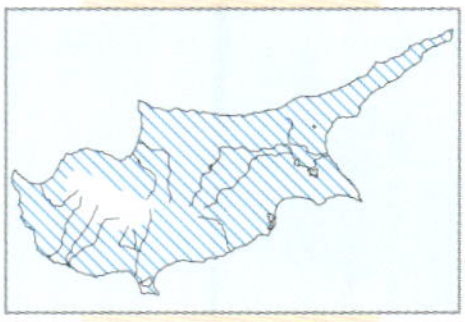

Garden Warbler *Sylvia borin* PM

L 14. Scarce to fairly common passage migrant in variable numbers. Spring passage Mar–Jun, peaking mid Apr–early May; fewer in autumn, mid Aug–early Nov, mainly Sep–early Oct. **Habitat** Light woodland with bushes and undergrowth; usually at low elevations. **ID** Uniform grey-brown warbler with square-ended tail, ***short relatively heavy bill***, practically no supercilium, often greyish neck-sides, ***rounded head and no white in tail or wings***. Most *Acrocephalus*, *Hippolais* and *Iduna* warblers have thinner, longer bills, *Acrocephalus* also have longer undertail-coverts and rounded tails. See juvenile Barred Warbler for separation. Usually keeps well hidden. **Voice** Alarm call a nasal, repeated *chet-chet-chet*, softer than Blackcap.

Barred Warbler *Sylvia nisoria* pm

L 15.5. Scarce passage migrant. Spring passage mid Mar–May; autumn passage late Jul–Oct, peaking mid–late Aug (when mainly in east). **Habitat** Varied scrub and bushy habitats, not montane forests. **ID** Large warbler, ***adult with crescentic barring on underparts*** (reduced in female) and ***yellowish iris***; fairly long tail ***has conspicuous white corners*** on landing; tips of greater coverts and tertials whitish. First-winter dark-eyed, recalling Garden Warbler, but ***larger, with white in tail, pale edges to wing-coverts and tertials, and some dark barring on undertail-coverts***; this plumage often remains until following spring. Young Eastern Orphean Warbler lacks pale markings on wing-coverts and tertials, and has contrasting dark ear-coverts, but may show dark marks on undertail. Usually furtive, keeping hidden; movements rather heavy. **Voice** Hard sparrow-like rattling *trrrrr* or *che-er-er-er-er-er-er*, distinctive, often interspersed in song or used as alarm.

Eastern Orphean Warbler *Sylvia crassirostris* PM

L 15. Fairly common passage migrant. In spring mainly Mar–May, peaking early Apr; autumn passage late Jul–early Oct (from late Jun in east), peaking mid Aug–early Sep. **Habitat** Bushy garrigue, including scrub and gorse; also orchards, olive and almond groves, usually near coasts or in foothills. **ID** ***Distinctly larger than Sardinian Warbler with grey crown contrasting poorly with blackish ear-coverts, and larger bill; adult has yellowish iris*** (orange in adult Sardinian, which also has reddish eye-ring), but many singing males have dark, mud-coloured irides. Tail-sides white. First-winter and younger females also ***lack pale iris***; former, like some first-summer males, recall large, sluggish Lesser Whitethroat with contrasting dark lores and ear-coverts, and dark markings on undertail-coverts. Often remains in cover. **Voice** Loud, varied, musical song recalls Common Nightingale, given from dense cover. Calls include a loud Lesser Whitethroat-like *tak* and a *trrr*.

Asian Desert Warbler *Sylvia nana* v

L 11.5. Vagrant. Occurs mid Nov–early Apr. **Habitat** Scattered bushes and trees; low scrub and salt bushes. **ID** Small, relatively long-tailed, sandy grey-brown warbler with ***rufous-brown tertials, rump and closed uppertail; spread tail tricoloured (dark brown with white at sides and rufous centre)*** often flicked or half-cocked; ***iris and legs yellowish***, fine bill largely yellowish, sometimes with pale area around eye. Terrestrial; hops and scuttles, or flies low into cover, where remains hidden. **Voice** Call a dry, weak purring *drrrrrr*, descending and fading out; also *chrr-rrr* and high-pitched *che-che-che-che*.

♀
ad
♂
Garden Warbler
Eurasian Blackcap
♀
♂
Barred Warbler
1st-winter
♀ 1st-winter
♂
♀
1st-winter
Eastern Orphean Warbler
1st-winter
♂ spring
Asian Desert Warbler

Lesser Whitethroat *Sylvia curruca* PM

L 13.5. Very common passage migrant. Occurs mid Feb–May, peaking late Mar–mid Apr; fewer late Jul–early Dec, peaking mid Aug–early Nov. Occasional in other months. **Habitat** Trees and bushes, often in montane forest. **ID** Race *curruca* occurs in Cyprus. Grey-brown above, browner wings, mid-grey crown and darker ear-coverts, variable in prominence; contrast between white throat and dusky-washed breast. The only *Sylvia* warbler on Cyprus with ***all-black legs***. First-winter Lesser Whitethroat has slightly paler upperparts and often an indistinct whitish supercilium. **Voice** Located by call, a rather short and hard *tek* often repeated; scolding alarm *jeh-jeh-jeh-jeh-jeh*.

Common Whitethroat *Sylvia communis* PM

L 14. Common passage migrant. *S. c. communis* regular; some *S. c. icterops* in spring. Occurs late Feb–early May, mainly late Mar–Apr; less frequent Aug–Nov, mainly mid Aug–early Oct. **Habitat** Scrub, bushes, broadleaf woodland and dense vegetation. **ID** Medium-sized, ***chestnut fringes to coverts and secondaries***, slightly longer tail, ***orangey-coloured legs and white eye-ring***. Female and first-winter lack male's grey-washed head and have dark (not orange) iris. *S. c. icterops* has brownish-fringed coverts and secondaries, resembling Lesser Whitethroat, but larger. **Voice** Migrants regularly call and sing. Song a fairly short, scratchy warbling outburst from perch atop bush or in display flight. Call a harsh *whet-whet-whet*; loud scolding alarm a repeated drawn-out *jaairh*.

Spectacled Warbler *Sylvia conspicillata* RB, mb?

L 12. Fairly common resident and possible breeding visitor. Mostly sedentary, but some move to lower ground or leave island mid Aug–mid Mar. Breeds late Jan–Jun. **Habitat** Open areas including salt flats, dry scrub and low bushes, from sea level to treeline, where thinly distributed. **ID** Resembles small, bright Common Whitethroat with ***conspicuous bright rufous wing-coverts and blackish-centred tertials broadly edged rufous***. Adult male has ***darker grey head*** (feathers tipped buff-brown when fresh) merging with ***blackish lores and around eye; pale eye-ring obvious***. Underparts pinkish with demarcated white throat. Fine, bicoloured bill, nearly orange legs. Female and first-winter duller, with brownish head and off-white throat. Moves inconspicuously in cover. **Voice** Song, given in flight or from open perch, usually starts with a few clear notes, followed by chatter with short, high-pitched phrases at fast tempo. Calls include quiet, rattled, soft Eurasian Wren-like *drrrrr* and short *tec*.

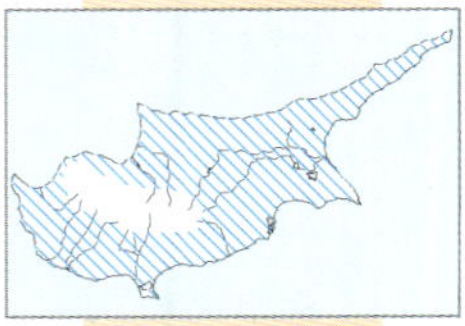

Subalpine Warbler *Sylvia cantillans* PM

L 12. Fairly common spring passage migrant, rare in autumn. Variable numbers, Mar–early May, peaking mid Mar–early Apr; rather scarce mid Aug–early Oct. **Habitat** Scrub and bushy areas; shuns upland forest. **ID** *S. c. albistriata* ('Eastern Subalpine Warbler' occurs in Cyprus) male ***bluish-grey above*** with browner wings and ***red orbital ring, dark brick-red throat and breast*** clearly demarcated from whitish belly and broad ***white moustachial stripe***; tail dark grey with white outer feathers. Some females like dull males, others (including first-winter) greyer, resembling Lesser Whitethroat but smaller with paler ear-coverts, yellow-brown legs and ***pale eye-ring***; may be confused with first-winter Spectacled Warbler. From female Sardinian Warbler by smaller size, finer bill, paler plumage with warm buff sides to throat and breast, and pale eye-ring. Usually skulks in dense cover. **Voice** Call a single dry *tek* sometimes repeated in series.

Ménétries's Warbler *Sylvia mystacea* V

L 12.5. Vagrant. **ID** Superficially like male Sardinian Warbler but slightly smaller and shorter-tailed, ***characteristically waves tail sideways or up and down. Male has black forehead and ear-coverts*** merging into grey crown and grey-brown mantle, wings with sandy feather fringes and black alula. In fresh plumage white moustachial stripe contrasts with salmon-pink throat, paler on breast, but whitish below when abraded. Dark, broad-based tail less rounded than Sardinian Warbler with white in outer feathers; ***eye-ring salmon-pink to red. Female paler and plain***; from Sardinian by paler upperparts, ***uniform buffish-white underparts*** (more contrasting white throat and dull brown flanks in Sardinian); eye-ring yellow-brown in adult female and first-winter. Mostly in low scrub or on ground. **Voice** Calls include Sardinian-like *chak* or *tret* or distinctive ***hard, quick-fire staccato rattle*** softer than Sardinian Warbler; sometimes also a softer *tshshshshsh*.

ad
1st-winter
Lesser Whitethroat
♀
♂
communis
icterops
♀
Common Whitethroat
♂ 1st-winter
♂
♂
♂ 1st-winter
♀
Spectacled Warbler
♀ 1st-winter
♂
♀
♀ 1st-winter
Subalpine Warbler
♂
♂ 1st-winter
♀
♀ 1st-winter
Ménétries's Warbler

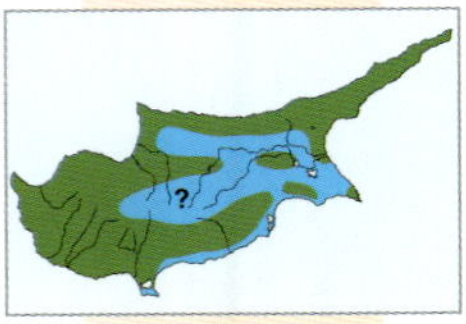

Sardinian Warbler *Sylvia melanocephala* RB, PM, WV

L 13. Very common to abundant resident, fairly common passage migrant and winter visitor. Spring mainly Mar–mid Apr; autumn passage from mid Aug, wintering Nov–Mar. First bred 1980s, and has rapidly expanded since. Breeds late Jan–Aug (double-brooded). **Habitat** Bushes and trees in most habitats at all altitudes, but rather scarce in dense forests of the Troodos. **ID** Male has *black head, red eye-ring and contrasting white throat*. Rare males have *all-black hoods*. Female and first-winter have rather dark brown upperparts; from female Subalpine by size, longer, more rounded tail, *longer bill, whitish throat contrasting with dull brown flanks* and inconspicuous pale eye-ring (red in adult female Sardinian). Rather skulking. **Voice** Song of hard call-like rattling notes mixed with short whistles at fast tempo, often in flight. Calls include a *sudden mechanical rattle* of 4–6 syllables, *chret-tret-tret-tret-tret* or slower *terit-terit-terit-terit*, also single loud, hard *tche*.

Rüppell's Warbler *Sylvia rueppelli* PM

L 14. Fairly common in spring, rare in autumn. Variable numbers, from late Feb–early May, mainly mid Mar–mid Apr; rare mid Aug–Oct. More common than Subalpine Warbler. **Habitat** Scrub and low thorny bushes, often near coasts. **ID** *Fairly large*, long-tailed warbler with *slow movements*. Upperparts greyish in all plumages (especially rump, contrasting with blackish tail), wings darker with *conspicuous pale fringes to tertials and greater coverts*; eye-ring and orbital ring brown (young) to reddish (adult), never whitish. Legs red-brown. Adult male has *black hood and throat separated by striking white moustachial stripe*. Adult female similar but forehead and throat usually mottled black and 'moustache' less clear-cut. **Voice** Calls include distinctive dry chatter or single clicked note.

Cyprus Warbler *Sylvia melanothorax* RB, MB

L 13. Scarce to fairly common resident and summer visitor. Endemic breeder. Currently declining in western areas (including Paphos District), increasing in parts of east. Small numbers depart island Oct–Mar. Breeds Feb–May. **Habitat** Scrubby and gorse-covered plains and hillsides, garden edges and lightly wooded areas, scarce above 1,300m. Rare or absent on hot inland plains. **ID** Resembles Sardinian Warbler but note *variable black spots or blotches on underparts including undertail-coverts; upperparts show obvious pale fringes to inner wing feathers* in fresh plumage (similar to Rüppell's, almost absent in Sardinian), and *bill distinctly bicoloured with pink base*. Adult male unmistakable: *black hood*, ash-grey mantle, *white moustachial stripe, black spots below*, with reddish-brown eyes and orbital ring, and legs warm straw-coloured. *Adult female and first-winter male have mottled black crown*, grey-brown upperparts and *mottled dark grey throat and breast with faint whitish moustachial stripe*; legs reddish-brown. First-winter female has crown uniform grey-brown lacking dark spots, and underparts with few or no spots. More confiding than Sardinian Warbler. **Voice** Jerky song much like Sardinian Warbler, but with infrequent whistles; rattling call like Sardinian, also a distinct bunting-like *pwit* (recalls alarm note of Cetti's Warbler), not heard from Sardinian.

Goldcrest *Regulus regulus* pm, wv

L 9. Uncommon to scarce passage migrant and winter visitor. Occurs Nov–Apr; mainly mid Nov–Dec, scarce midwinter and spring. **Habitat** Mountain cedars and pines, occasionally lower elevations. **ID** Small and compact, with short tail and fine bill; *two wingbars*, yellow central crown-stripe (orange-red in male but concealed by yellow feathers). *Dark eye in pale face, lacks white supercilium and dark eyestripe of Firecrest*. Very active with fluttering movements, often foraging with tits and Chiffchaffs; hovers frequently. Often fearless. **Voice** Calls almost constantly, a very high, Coal Tit-like *see-see-see* or *sit-sit*.

Firecrest *Regulus ignicapilla* v

L 9. Vagrant. **Habitat** As Goldcrest, but often in undergrowth. **ID** Size, structure and behaviour as Goldcrest, but brighter with paler green upperparts, tinged bronzy on shoulders/neck-sides, and frosty-white below. Head pattern striking: broad white supercilium contrasting with distinctive black eyestripe. Has yellow (female) or *fiery-orange* (male) central crown-stripe. **Voice** Many calls similar to Goldcrest, but more sibilant in others; sharp *iizt*; also a tit-like, piping *peep*.

♂
♀
♂ 1st-winter
Sardinian Warbler
♂
♀
♀ 1st-winter
Rüppell's Warbler
♀
♂ 1st-winter
Cyprus Warbler
♂
♀ 1st-winter
juv
♂
♀
Goldcrest
juv
♂
♀
Firecrest

Eurasian Wren *Troglodytes troglodytes* RB, wv?

L 9.5. Fairly common breeding resident. Some residents, and possibly winter visitors, occur away from forests mid Sep–mid Mar. Breeds Feb–Jun. **Habitat** Mainly bushy montane valleys with streams, in Troodos and Kyrenia ranges; also dense tangles in riverine forests and ravines at low elevation. **ID** Regional endemic *T. t. cypriotes* ('Cyprus Wren') ***has longer bill and more barring below*** than Western European races. ***Tiny and short-tailed, tail characteristically held cocked***; warm brown plumage. Reminiscent of small mouse when moving, often on ground. Bill thin, supercilium pale, with wings, tail and flanks barred; often remains hidden or seen in low, rapid, whirring flight between cover. Calls insistently, usually with 'attitude', but often forages silently. **Voice** Musical song, *unexpectedly loud* for such a small bird, of very fast trills with different pitch and tempo, lasting 4–6 seconds. Calls include a hard dry *tuk*, or often a rattling *trurrrr*; reminiscent of sound of a fisherman's reel.

Wallcreeper *Tichodroma muraria* wv, pm?

L 17. Uncommon winter visitor and possible passage migrant. Occurs late Oct–early Apr, peaking Nov–Feb. Occasional passage early–mid Mar. **Habitat** Local. Ravines and gorges with running water, castle walls and south coast sea cliffs. **ID** Unmistakable and spectacular. Grey above, male has black throat in summer, with short dark tail and ***long, thin decurved bill; habitually creeps on rocks and cliffs often flicking wings to show flashes of red***. In flight, ***short tail, rounded black-and-red wings with white primary spots***, recalling small Hoopoe, with butterfly-like wingbeats. **Voice** Song a clear, fluty, rising whistle, rather loud and melodious.

Short-toed Treecreeper *Certhia brachydactyla* RB

L 13. Common breeding resident. Breeds Mar–Jun. Some movement outside breeding season, away from dense forests to more open areas. **Habitat** Coniferous and mixed forest, from near sea level to highest summits. Nests in cracks in trees, eaves and gutters, occasionally in rock crevices. **ID** Endemic race *C. b. dorotheae* ('Dorothy's Treecreeper') small with long decurved bill, streaky brown cap and upperparts, rather indistinct white supercilium fading above lores. Dark bar on secondaries of even width. This race slightly greyer above and on rump, whiter below, and paler grey on flanks. Often at high density. Regularly perches across thin branch in treetop. **Voice** Short and simple with irregular rhythm, a curl in the middle, and last note rising in pitch, *tu-ti-teroi-e-tit*; distinctive once learned. Call a ***shrill Coal Tit-like teeut*** or *tuu*, like first note of song, uttered in series of variable length and pitch.

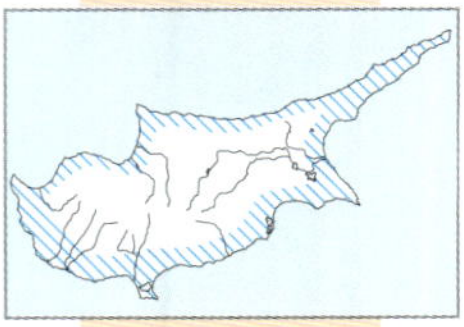

Rose-coloured Starling *Pastor roseus* pm

L 21. Scarce and irregular passage migrant: Apr–May, peaking mid May; mostly immatures Aug–early Sep; occasional other months. Usually singly. **Habitat** Often near coast, orchards and olive groves; perches prominently. **ID** ***Black-and-pink adult male unmistakable; crest raised when singing***; female more sombre. From young Common Starling by ***shorter yellow bill with curved culmen, pale lores, paler grey-brown upperparts, whitish-grey rump and dirty whitish underparts***. **Voice** Migrants usually silent. Calls resemble Common Starling, rapid, high-pitched, musical chatter. [Alt: Rosy Starling]

Common Starling *Sturnus vulgaris* WV, PM

L 21. Common winter visitor and passage migrant. Rather local with influxes and unpredictable movements. Occurs Oct–early Mar, with ones and twos in other months. **Habitat** Cities and towns; inland plains, agricultural areas, cereal fields and wetlands. Groups often seen late afternoon heading to roost. **ID** ***Short tail and short, pointed, triangular wings, fast, straight flight*** with characteristic frequent glides or brief wing closures. ***Breeding adult has glossy green blackish plumage with numerous tiny white spots***, though spots almost absent in spring in some races that occur on Cyprus, having bluish-green on mantle and bronzy sheen on head. ***Juvenile drab brown, moulting to blackish with prominent white spots in first-winter***, resembling winter adult. Feeds aerially or on ground; walks quickly. **Voice** Flight call a short buzzing *tcheer*, alarm a hard *kjet*.

ad
Eurasian Wren
♂ summer
winter
Short-toed Treecreeper
Wallcreeper
juv
♂ summer
Rose-coloured Starling
juv
ad winter
ad summer
Common Starling

PLATE 75: THRUSHES

Ring Ouzel *Turdus torquatus* v

L 24. Vagrant or rare migrant. Late Sep–early Mar, mainly Nov. **Habitat** Forest edges, mainly montane. **ID** Resembles Eurasian Blackbird; male has *white breast crescent and pale fringes to wing feathers*, appearing 'frosty-winged' in flight. Female dark brown with *buffish breast crescent* (absent in first-winter female), *pale fringes to wing feathers and underparts, giving scaly appearance*. Shy. **Voice** Hard scolding *tek-tek-tek*, and shrill chatter.

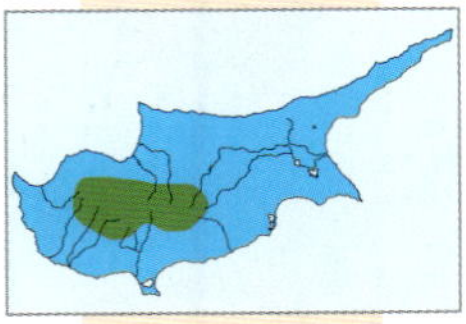

Eurasian Blackbird *Turdus merula* RB, PM, WV

L 25. Fairly common resident, passage migrant and winter visitor. Migrants late Oct–Mar, occasionally later. Usually shy. Breeds Apr–Jun (has bred since 1980s, increasing). **Habitat** Bushy areas. Breeds in pine and mixed forests. **ID** Male black with yellow bill. Female dark brown with dark bill; slightly paler throat, dark mottled breast. Juvenile similar but with more breast mottling. Feeds on berries. **Voice** Fine songster with rich flute-like notes. Calls include *chak, chak, chak*, accelerating into fast outburst; a thin *tsee* in flight.

Dusky Thrush *Turdus eunomus* v

L 24. Vagrant. **ID** Upperparts olive-brown, closed wing with rufous panel; bold white supercilium, black lores and patch on ear-coverts, white throat and one or two boldly black-flecked breast-bands, extending onto flanks; in non-breeder breast-bands indistinct and upperparts paler; tail dark brown; underwing-coverts rufous. Bill base bright yellow. **Voice** Flight calls include *srrii-i*; also *kwae-waeg* or *tjshah-tjshah*.

Naumann's Thrush *Turdus naumanni* v

L 24. Vagrant. **ID** Orange wash to face (winter adult, absent in first-winter), orange-blotched underparts and rusty-orange tail are key features. **Voice** As Dusky Thrush.

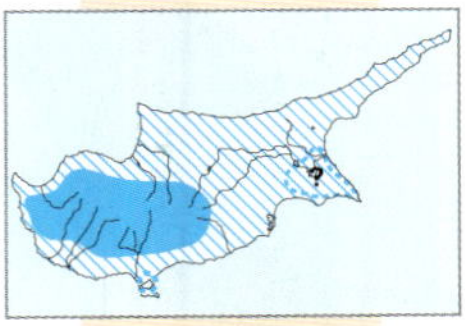

Fieldfare *Turdus pilaris* pm, wv

L 25. Uncommon passage migrant and winter visitor. Large groups in some winters. Occurs mid Oct–early Apr, spring mid Feb–mid Mar; autumn Nov–mid Dec. **Habitat** Montane forests; almond groves at lower altitudes. **ID** Heavily built, with *rufous-brown mantle and coverts, contrasting with pale ash-grey head and rump; black tail, rusty-yellow breast boldly streaked/spotted blackish, flanks with dense 'V'-shaped marks and underwing-coverts white*; young duller. Flight slightly undulating. **Voice** Loud *dscach-dscach-dscach*; thin *veid* contact note.

Redwing *Turdus iliacus* wv, pm

L 20. Rare to uncommon winter visitor and passage migrant. Less than annual. Occurs Nov–Feb, rare Mar–early Apr; autumn migration light or absent, peaking mid Nov. **Habitat** Juniper and pine forests at all altitudes; also gorse and broom. **ID** Smaller than Song Thrush, with *prominent buff-white supercilium and submoustachial streak, streaked underparts, rusty-red flanks and underwing-coverts* (hard to see); *also voice*. Flight fast and straight. **Voice** Call a thin, metallic, drawn-out *steeef* or *zeeip*.

Song Thrush *Turdus philomelos* WV, PM

L 22. Common winter visitor and passage migrant. Occurs mid Oct–mid Apr. Influx late Dec–Feb, most depart early Apr. Summer sightings, but breeding not proven. **Habitat** Forests, gorse, brambles and olive groves. **ID** Small brown thrush, with *blackish spots on whitish underparts*; underwing-coverts *buff*. Flight fast and straight. Alone or in small groups, shy. **Voice** Song (heard from migrants in Feb–Mar) mix of fluty and shrill notes, repeated 2–4 times, *di-du-weet, di-du-weet, di-du-weet; – dwi-dwi-dwi; – du-drid-du-drid; – peeoo-peeoo-peeoo-peeoo*.... Call a sharp *zit* or *zip*.

Mistle Thrush *Turdus viscivorus* wv, pm

L 28. Uncommon winter visitor and passage migrant. Occurs late Oct–Mar; mainly mid Nov–mid Dec. Uncommon in spring. **Habitat** Montane forests; bushy areas on lower ground. **ID** Resembles large Song Thrush. White underwing-coverts conspicuous in flight; *grey-brown upperparts, white corners to uppertail* (visible on landing), *whitish underparts evenly spotted blackish*. Pot-bellied, with upright stance. Often shy. **Voice** Flight call a loud, dry churring *trrrrrr*, delivered repeatedly, vehemently in alarm.

♂
Ring Ouzel
♀
♂
♀
Eurasian Blackbird
1st-winter
ad
ad
Dusky Thrush
Naumann's Thrush
ad
ad
Fieldfare
Redwing
ad
ad
Mistle Thrush
Song Thrush

Rufous-tailed Scrub Robin *Cercotrichas galactotes* pm

L 15. Rare passage migrant. Occurs late Mar–mid May, peaking mid Apr; in autumn mid Aug–early Oct, mainly late Aug–early Sep. **Habitat** Scrub and bushy areas. **ID** Skulking; slightly smaller and slimmer than Common Nightingale, with ***long tail often cocked and spread***. Upperparts grey-brown in *C. g. syriaca* contrasting with ***rufous rump and tail***, the latter with ***prominent black subterminal band and white tips*** above and below; distinctive white supercilium contrasts with blackish eyestripe. In *C. g. galactotes* upperparts rufous, concolorous with rump and tail. Not shy but usually near cover; feeds on ground. **Voice** Calls include a hard *teck*, a low rolling *schrrr* and sibilant drawn-out *iiiip*; a distinctive penetrating *ssweep* when agitated. [Alt: Rufous Bushchat.]

European Robin *Erithacus rubecula* PM, WV

L 14. Very common passage migrant and winter visitor. Earliest late Aug, main passage mid Oct–early Dec; most depart by Mar, stragglers late Apr. Migrants often sing, but breeding unknown. **Habitat** Trees and bushes at all elevations. **ID** Plump, upright with ***diagnostic orange-red face and breast***, large dark eyes and short wings. Usually hops on or close to ground, shy in Cyprus; voice distinctive. **Voice** Song short but crystal clear and plaintive with abrupt changes in pitch and tempo, often starts very high-pitched, followed by a much lower-pitched, sometimes bubbling, series. Usual call a sharp clicking *tic* or *tic-ik*, often repeated; also a thin drawn-out *tseer*.

Bluethroat *Luscinia svecica* PM, WV

L 14. Common passage migrant and winter visitor. Occasional passage from Aug, mainly Oct–early Dec; winter visitors depart mid Feb; variable spring passage Mar–Apr. *L. s. svecica* ('Red-spotted Bluethroat') predominates on passage and in winter; *L. s. cyanecula* ('White-spotted Bluethroat') uncommon, peaking late Oct; *L. s. magna* ('Caucasian Bluethroat') rare. **Habitat** Reedbeds, marshes, ditches, irrigated fields and riverine vegetation. Usually at low elevations. **ID** Size and shape resemble European Robin but slimmer, longer-legged and darker above; generally more skulking. ***Rust-red sides to tail base diagnostic***, obvious in flight. Breeding male has ***blue chin and throat*** framed by black and rusty bands on breast. Racial variation complicated and only adult males are confidently assigned: throat patch red in *svecica*, white in *cyanecula* (note: many *svecica* appear 'white-spotted' in winter, but on closer inspection show reddish tone to lower edge of central spot); entirely blue in *luristanica*. Most non-breeders have pale throat, ***black malar stripe connected to dark necklace*** and variable amount of blue or red admixed. Curious birds occasionally perch high; but mainly slinks around tail cocked, before darting back into cover. **Voice** Calls include a hard *tack*, soft *hweet*, an odd *dwzeer* and repeated 'snipping' note.

White-throated Robin *Irania gutturalis* v

L 16. Vagrant. Occurs mid Mar–early May; rarely autumn. **Habitat** Dense scrub and woodland; usually on headlands. **ID** Size and movements recall nightingales. Striking male has ***black sides to face and head framing white central throat, rusty-red underparts***, blue-grey upperparts, ***black tail*** and whitish supercilium. Scarcer variant is paler orange below and has black line below white of chin. Grey-brown female has ***dark brown tail, ochre-buff sides to body***, and whitish throat bordered grey-buff on sides of head and breast. Bill long. Skulks, often on ground, wings lowered. **Voice** Calls include a nightingale-like *kerr-r-rr-rr*, wagtail-like *tzi-lit* and a hard *tack* in alarm.

Red-flanked Bluetail *Tarsiger cyanurus* v

L 14. Vagrant. **ID** Behaviour recalls European Robin, remaining close to cover. Male blue above, with brighter ***blue shoulders and uppertail-coverts, dull bluish tail and orange-buff flanks***; flight feathers fringed bluish. First-year and female have some bluish on uppertail-coverts and base of outer tail feathers (hard to see), ***white eye-ring and narrow white throat*** bordered by olive-grey sides to throat and breast, and ***orange flanks***, often concealed. Legs dark. Jerks tail unlike shivering of redstarts. **Voice** Hard *tsak* or short series of quiet *tck, tck, tck* notes; also a frog-like note.

galactotes
(Near East)
ad
syriaca
ad
Rufous-tailed Scrub Robin
ad
European Robin
♀
♂
magna
♂
svecica
♂
cyanecula
Bluethroat
♂ 1st-winter
♂ 1st-winter
♀
♂
White-throated Robin
♂
♀
Red-flanked Bluetail

PLATE 77: NIGHTINGALES AND REDSTARTS

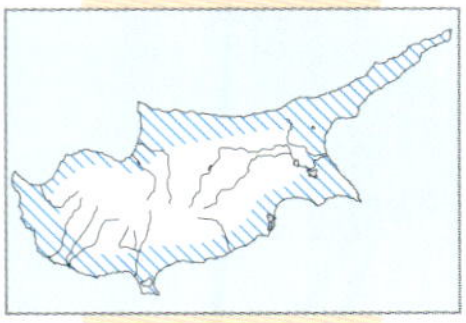

Thrush Nightingale *Luscinia luscinia* — pm

L 16. Uncommon passage migrant. Mainly late Mar–mid May, peaking mid–late Apr (5–10 days later than Common Nightingale); rare mid Aug–early Oct. **Habitat** Dense bushes and woodland; mostly on headlands and peninsulas. **ID** Dark olivaceous-brown, like a small thrush, with ***rusty-red tail*** and pale underparts, whitish throat and eye-ring. Very similar to Common Nightingale but ***darker brown above, with duller red tail and darker brownish-grey breast and flanks indistinctly mottled; far-carrying song distinctive***. Skulking. **Voice** Song, often at night, recalls Common Nightingale but louder; includes ***characteristic hard chuck notes***, dry rattles and clear whistles; lower-pitched with delivery more mechanical, less variable *djüllock – djüllock – djüllock – drllrllrllrllrll-pst*, lacking distinctive crescendo of Common Nightingale. Calls include high-pitched *hiiid* and dry, rolling rattle.

Common Nightingale *Luscinia megarhynchos* — PM, MB

L 16. Fairly common passage migrant and breeder. Spring mid Mar–early May, peaking early–mid Apr. Scarce in autumn, mid Sep–Oct. Breeds late Mar–Jun (apparently declining). **Habitat** Thickets, orchards, olive groves, lush river valleys and ravines on passage. Breeds in broadleaf woodland valleys in main ranges. **ID** Similar to Thrush Nightingale but ***more russet-brown upperparts, paler rusty-red tail***, and more conspicuous whiter eye-ring; underparts 'cleaner', ***lacking mottled impression on breast and flanks of Thrush Nightingale***. Juvenile spotted buff and brown above and on breast and flanks, pale-spotted median wing-coverts. Skulking, even when singing. **Voice** Beautiful song by day and night, high-pitched, comprising loud, rich, warbling whistles with ***distinctive crescendo***, *lu-lu-lu-lu-lee-lee* often a characteristic starting sequence, not heard in Thrush Nightingale; lacks strong chuck notes or frequent rattles. Calls similar to Thrush Nightingale, including *wwheep* and a soft frog-like croak, usually from dense cover and often heard on migration.

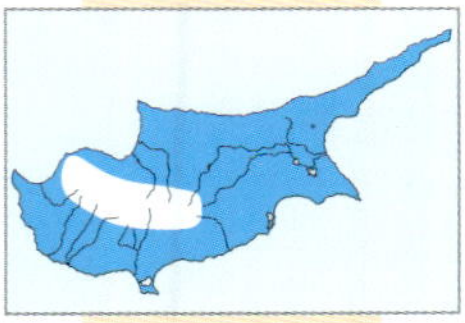

Black Redstart *Phoenicurus ochruros* — WV, PM

L 15. Common winter visitor and passage migrant. Occurs Sep–mid May. Autumn passage mid Oct–early Dec, wintering late Oct–Feb; spring passage Mar–early Apr, occasionally later. Wintering females outnumber males three to one. **Habitat** Rocky hillsides, open woodland, villages, built-up areas, coastal cliffs and archaeological sites. **ID** Race *gibraltariensis* occurs. Winter male ***grey above, all blackish below, some with white wing panel; female uniform, paler grey above and below, and lacks wing panel***. Female similar to Common Redstart but ***slightly darker and drabber, especially below*** (Common Redstart more buffish below, with warmer flanks and breast-sides). Often on ground; bobs body and shivers tail. **Voice** Alarm a dry *eet-tk-tk-tk*.

'Red-bellied' Black Redstarts

Fewer than ten reports on Cyprus, all in winter, of male 'red-bellied' Black Redstarts, race *phoenicuroides* ('Central Asian Black Redstart'). Also historic reports of *ochruros* (Turkey to Iran) and *semirufus* (Near East) and much confusion between these in the field. Male *semirufus* (not illustrated) has black upperparts and throat to ***lower breast***, sharply defined from deep red below. Male *phoenicuroides* greyer above and black below extends only to mid-breast; male *ochruros* is variable; some resemble *semirufus* or *phoenicuroides*, but many have ***black breast merging into reddish belly***. Some female *phoenicuroides* very similar to Common Redstart, but usually suffused rustier on lower breast and belly, with grey cast to upperparts, even if head and throat pale.

Common Redstart *Phoenicurus phoenicurus* — PM

L 15. Fairly common passage migrant. Spring Mar–early Jun, peaking early–mid Apr; autumn late Aug–mid Nov, peaking Oct. Race *samamisicus* 'Ehrenberg's Redstart' (breeds Turkey eastwards) is an uncommon spring passage migrant (very few in autumn). Peak passage mid Mar–early Apr (10–14 days earlier than nominate Common Redstart). **Habitat** Light woodland or open areas with scattered trees. **ID** Male has ***black cheeks and throat contrasting sharply with rusty-red breast*** and belly; crown grey with ***pronounced white forehead***; autumn male has black areas fringed pale. Male ***samamisicus has white wing panel*** and, often, darker upperparts. Female brownish-olive above; from similar Black Redstart by ***paler, warmer buffy-white underparts***. Hunts insects in flycatcher-like fashion; bobs body and shivers tail (as Black Redstart). **Voice** Call resembles Willow Warbler's soft *wheet*, often followed by *tuuk-tuuk*.

ad
ad
Common Nightingale
megarhynchos
Thrush Nightingale
♂
gibraltariensis
♂
♀
ochruros
♂
phoenicuroides
♂
phoenicurus
♀
♂
samamisicus
Common Redstart
♂ 1st-winter

Pied Bush Chat *Saxicola caprata* v

L 13. Vagrant. **ID** Slimmer and slightly longer-tailed than other stonechats; flicks tail less. ***Jet-black male has white belly, rump and narrow shoulder patch*** visible in low jerky flight. Female from other chats by ***unstreaked sooty earth-brown upperparts and breast***, creamy belly and rufous-orange rump, some rusty-brown on breast and slight supercilium. **Voice** Alarm a curt *chuk* and a *chek-chek-trweet*. [Alt: Pied Stonechat.]

Whinchat *Saxicola rubetra* PM

L 12.5. Common passage migrant. Occurs Mar–May, mainly Apr–early May; autumn passage mid Aug–Nov, mainly mid Sep–early Oct. **Habitat** Open country with bushes and other suitable perches. **ID** Short-tailed chat with slightly smaller head and longer primary projection than European Stonechat, reaching almost halfway down tail. From stonechats by ***streaked brownish rump and white sides to base of tail. Male has bold, white supercilium and white stripe between blackish head-sides and orangey throat***; female duller with paler throat and browner head-sides. ***White spot visible on primary-coverts***, especially in flight (absent in some first-winter birds). Confusion possible in autumn with first-winter Siberian Stonechat. **Voice** Silent on passage.

European Stonechat *Saxicola rubicola* WV, PM

L 12. Very common winter visitor and passage migrant. Occurs mid Sep–early May. Autumn passage occasionally late Sep, mainly mid Oct–early Nov, when winter visitors arrive. Departs from late Feb–early Mar, latest to mid Apr. Light spring passage early Mar, mainly in east. Usually winters in pairs. **Habitat** Coastal plains, open country and foothills. **ID** Short-tailed and short-winged with large rounded head and upright stance, frequently flicks wings and tail. Adult male easily separated from Whinchat by ***black head and throat, reddish breast, white neck patch and black tail***. Female duller with dark brown head and usually throat, reddish-brown upper breast, with Whinchat-like supercilium in autumn. Male has variable-width grey-brown to white rump; if latter usually (but not always) narrow or with dark streaks; ***grey underwing-coverts and axillaries*** (blackish in Siberian Stonechat) with white collar narrower and less extensive than male *maurus* Siberian Stonechat. Some male European Stonechats have an extensive white rump and long white collar. These 'Siberian'-looking individuals might be unique to southern populations of European Stonechat. Female has dark earth-brown back, usually a ***streaked brownish rump***; dusky or whitish throat (Siberian typically only whitish), but some indistinguishable from *maurus*. Female Siberian Stonechat has paler sandy to warm buff upperparts, often with a more obvious supercilium and a pale unstreaked rump. First-winter European Stonechat is darker, lacks pale rump of *maurus* and has less extensive white on coverts. **Voice** Call *wheet-trak-trak* like pebbles hit together.

Siberian Stonechat *Saxicola maurus* pm

L 12. Rare spring passage migrant, occasionally autumn. 'Siberian' races mid Feb–Apr, peaking late Mar; fewer autumn and winter. Females infrequent (or overlooked). **Habitat** Similar to European Stonechat, but most occur on peninsulas and headlands, usually near coasts. **ID** Breeding male has broader, more extensive white collar than European Stonechat and ***conspicuous, broad, white to salmon-pink or orange-buff rump***. First-winter Siberian Stonechat recalls autumn Whinchat and female lacks clear-cut, long, creamy-white supercilium of Whinchat, but difficult to separate from European Stonechat (see above). Males of different subspecies described below.

S. m. maurus (breeds European Russia) has longer primary projection than short-winged European Stonechat (of western and central Turkey); male *maurus* has ***unstreaked orange-buff to white rump*** (wider than European male), ***black tail***, broader white half-collar; ***paler back and jet-black axillaries*** (greyish-white in male European).

S. m. variegatus (formerly *S. m. armenicus*, breeds south-east Turkey and Iran) is also long-winged but has less white at sides of tail (not always visible) than *hemprichii*; male has dark chestnut breast contrasting with pure white belly.

S. m. hemprichii (Caspian Stonechat; formerly *S. m. variegatus*, breeds Caspian), most common of Siberian races occurring has even longer primary projection (approaching Whinchat); ***much white in sides of tail base*** (recalling Northern Wheatear); the palest subspecies, male has warm buff upperparts and large white patches on rump, sides of neck and shoulders.

♂
♀
1st-winter
Whinchat
Pied Bush Chat
♀
♂
♂
♀
European Stonechat
hemprichii
variegatus
♀
Siberian Stonechat
♂ winter
maurus
♂
♂ 1st-winter

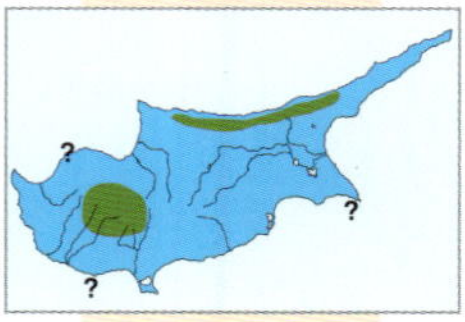

Blue Rock Thrush *Monticola solitarius* PM, WV, rb

L 21. Fairly common passage migrant, winter visitor and dispersive breeder. Autumn migration late Aug–early Dec, peaking Oct–Nov. Spring passage Mar–mid May. Breeds Jan–Jun. **Habitat** In winter on coastal cliffs, dam walls, escarpments, boulders and crags; also abandoned villages and rooftops. Nests in holes in crags, quarries and cliffs. **ID** Slightly larger and longer-tailed than Common Rock Thrush, with longer bill. Male has ***dull inky-blue plumage*** (looks black at distance); in winter can show fine buffish fringes. Female dark brown, resembles female Eurasian Blackbird but has longer bill, shorter tail and dull buff spotting and barring on underparts; some show bluish tinge to upperparts. Fairly shy but will perch in full view, remaining quite still. **Voice** Blackbird-like far-carrying song, melancholic, short, fluty phrases interspersed by long pauses. Sings in winter quarters. Calls include a hard *chak* and high *tsee*.

Common Rock Thrush *Monticola saxatilis* pm

L 19. Uncommon passage migrant. Occurs late Feb–May, peaking late Mar–mid Apr; scarce late Aug–mid Nov or later. Occasionally seen in mountains Jun–Jul, but breeding not reported. **Habitat** Mainly foothills with scattered boulders and rocky outcrops; almost any habitat on passage. **ID** Rather small and short-tailed with longish pointed bill. Adult male easily identified; in winter entire plumage appears scalloped. Female and first-winter browner, similar to winter male, resembling female Blue Rock Thrush but have ***pale spotting on upperparts, rusty tail, and scaling on warmer buffier underparts***. Often shy and elusive. **Voice** Song similar to Blue Rock Thrush but less melancholic. Calls include a loud *chak*. [Alt: Rufous-tailed Rock Thrush.]

Spotted Flycatcher *Muscicapa striata* PM, mb

L 14. Common passage migrant and localised breeding visitor. Occurs Mar–Nov. Spring passage Mar–May, mainly late Apr–early May; in autumn, late Jul–mid Nov, mainly Sep–early Oct. *M. s. neumanni* breeds mid Apr–Jul. **Habitat** Any area of woodland or scattered trees on passage; nests on sheltered slopes in montane forests, and mixed woodland in river valleys, including at lower elevations. **ID** Brownish-grey, long-winged flycatcher with ***streaked forehead, crown and breast*** (not always obvious), ***variable pale edges to greater coverts and tertials*** (depending on subspecies and wear), and ***blackish bill*** and legs. Faint whitish eye-ring. Sexes similar. *M. s. neumanni* is distinctly paler and greyer above than nominate *striata* (which occurs on passage). Makes short aerobatic flights (may hover) from exposed branch to catch prey, often returning to same perch. Perches upright, often flicking wings. **Voice** Call a sharp *tzeet*; alarm *isst-tek*.

Red-breasted Flycatcher *Ficedula parva* pm

L 12. Rare passage migrant. Occurs Apr–mid May and Sep–mid Nov. **Habitat** On passage anywhere with trees or scrub. **ID** Small flycatcher with conspicuous white patches at sides of tail base, whitish eye-ring, and straw or yellowish base to bill (when seen well); adult male has ***reddish-orange throat and sometimes upper breast***, and lead-grey sides to head and neck. (Migrants unlikely to be seen in breeding plumage, especially in autumn.) Female and second-year male have buffish-white throat. Often cocks tail. **Voice** Call a dry, rolling *terrrr*; also a thin *tsri*, *tek* or, in alarm, *tee-lu*.

♂
♀
Blue Rock Thrush
♂
Common Rock Thrush
♀
1st-winter
ad
♂ 1st-winter
1st-winter
♂
neumanni
Spotted Flycatcher
Red-breasted Flycatcher

Eurasian Pied Flycatcher *Ficedula hypoleuca* PM

L 13. Fairly common spring passage migrant, rare in autumn. Occurs in very variable numbers mid Mar–May, peaking mid–late Apr; in autumn Sep–mid Oct. **Habitat** Trees and woodland. **ID** Male black and white with *black hindneck*, small white forehead spot (often divided in centre) and *narrow white streak at base of primaries* (sometimes absent). Some males (rare) are grey-brown above but still have white spot on forehead. First-winter has dark tertials, white-bordered outer webs as Collared Flycatcher, with blob at end and only *small white streak at base of primaries*. Separation of female Pied Flycatcher from Collared and Semi-collared Flycatchers covered under those species. Like other black-and-white *Ficedula*, often flicks wings. Easily confused with Semi-collared Flycatcher, especially in autumn. **Voice** Call on passage a *short metallic twink*, very different from both Collared and Semi-collared.

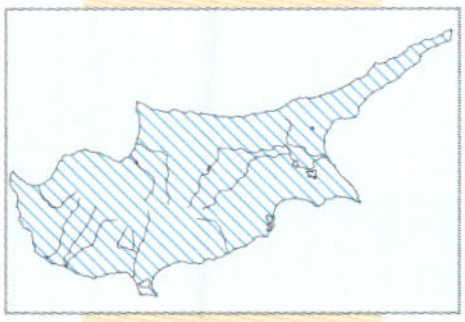

Collared Flycatcher *Ficedula albicollis* PM

L 13. Fairly common spring passage migrant, rare in autumn. Occurs in variable numbers; scarce some years, usually common, mid Mar–May, peaking mid–late Apr (on average a few days earlier than Eurasian Pied Flycatcher); in autumn Sep–early Oct. **Habitat** Trees and woodland. **ID** Male from Eurasian Pied and Semi-collared Flycatchers by *complete white collar on hindneck*; more white in wings than Pied Flycatcher, including *large white spot at base of primaries* (at most thin white streak in Pied), larger white spot on forehead and, often, paler rump. Female from similar Pied Flycatcher by *usually greyer, less brownish upperparts*, typically pale grey hindneck, *often broad white spot at primary bases, and call note*; sometimes inseparable from female Semi-collared Flycatcher (which see), unless hindneck is paler greyish, although broad white edges to darker tertials form clear wedge in female and first-winter Collared Flycatcher. **Voice** Call a thin piping *eeet*, very different from Pied Flycatcher.

Semi-collared Flycatcher *Ficedula semitorquata* pm

L 13. Uncommon spring passage migrant. Occurs in variable numbers, mid Mar–mid May, peaking early–mid Apr; rare in autumn when difficult to identify. The earliest of the *Ficedula* in spring. **Habitat** Trees and woodland. **ID** Both sexes resemble cross between Collared and Eurasian Pied Flycatchers. Male from Pied Flycatcher by *white half-collar on sides of neck, larger white spot at base of primaries* (absent, or just a narrow streak in Pied), more white on sides of tail and white-tipped median coverts (rare in Pied); extent of white in rest of wing and on forehead intermediate between Collared and Pied. Female resembles Collared Flycatcher, i.e. greyer above than Pied; *often* (but not always) *more white at primary bases and on tips of median coverts (as well as greater coverts)*; however, second wingbar on median coverts is not reliable for first-winters. **Voice** Call single dry *thuk* (like Red-breasted Flycatcher), also hard *tack*, alarm resembles Collared Flycatcher's *eeet*; song closer to Pied Flycatcher but higher pitched and simpler.

White-throated Dipper *Cinclus cinclus* FB

L 18. Former breeder and not recorded since 1945. **Habitat** Swift-flowing clear hill streams; nests beside running water, in crevice behind waterfall, under bridge or in bank. **ID** Stout, blackish-brown with *short, often cocked tail and large white 'bib'*, typically seen on a rock midstream or zipping away low over water. Juvenile greyish above, with grey scales or bars on whitish underparts. *Plunges or walks into water, swimming on surface or underwater*. Flight low over water, rapid and straight, following stream or river course. Generally solitary. **Voice** Song scratchy and squeaky with liquid warbling, heard even in winter; call distinctive, a penetrating metallic *zrit*.

1st-winter
1st-winter
♀ spring
♀ spring
♂ spring
♂ spring
♂ variant
Eurasian Pied Flycatcher
Collared Flycatcher
1st-winter
♀ spring
♂ spring
Semi-collared Flycatcher
juv
ad
White-throated Dipper

Northern Wheatear *Oenanthe oenanthe* PM

L 15. Very common spring passage migrant, less so in autumn. Spring passage mid Feb–May (up to ten days later than Isabelline Wheatear), mainly early Mar–mid Apr, peaking early Apr; autumn mid Aug–early Dec, mainly mid Sep–Oct, peaking late Sep. **Habitat** Coastal headlands, beaches and foothills; farmland, ploughed or cut fields and stony slopes, often with bushes. **ID** Relatively short-tailed wheatear with ***blackish terminal tail-band of even width***. Male has ash-grey crown and back, white supercilium, thin black eyestripe and black ear-coverts. Female greyer-brown than similar Black-eared Wheatear; first-winter ***best identified by shorter tail*** (wingtips reach closer to tail tip) and ***tail-band of even width***. In Black-eared tail longer, wings extend halfway or less along tail, which shows more white and narrower black band, with black extending upwards on outer tail feathers. Restless; bobs body, wags tail and flicks wings. **Voice** Alarm a hard *tack-tack* or *hiid, tack-tack*.

Isabelline Wheatear *Oenanthe isabellina* PM

L 16. Very common spring passage migrant, less so in autumn. Spring passage Feb–May (or earlier), peaking mid–late Mar, declining from mid Apr; in autumn late Jul–mid Nov, mainly mid Aug–Oct, occasionally to Dec. **Habitat** Coasts and foothills, favours barren, dry areas, sandy foreshores, ploughed fields and wasteland. **ID** ***Resembles large female Northern Wheatear, usually more robust***, with longer legs and apparent bill, slightly shorter tail and more upright stance. ***Best separated by isolated black alula in paler, more uniform sandy wings, contrasting less with upperparts***, broader pale wing feather fringes than Northern Wheatear (some first-winter Northern have pronounced dark alula); ***broader black terminal tail-band with shorter stalk to 'T'. Supercilium usually broadest and whiter in front of eye*** (in Northern narrower and buffier in front of eye). In low flight shows ***half-translucent, dark-tipped primaries; underwing-coverts and axillaries buffish-white***. Runs mostly; wags tail strongly and frequently. **Voice** Alarm call *tjack-tjack*, sometimes followed by slightly descending *hiu* or *diu*.

Desert Wheatear *Oenanthe deserti* pm

L 14.5. Uncommon spring passage migrant, rare in autumn. Spring mainly Mar–mid Apr, occasionally earlier, peaking early–mid Mar; autumn Sep–mid Nov. Mostly single males. Has occurred in winter. **Habitat** Coastal sandy scrub, ploughed fields, salt flats and dusty marginal land. **ID** In flight, from all other wheatears by ***almost all-black tail*** (no white at sides). Male recalls black-throated form of Black-eared Wheatear, but ***black throat joins narrower black of wings*** (often contrasting with whitish scapulars); in flight, whitish wing panel and white ***rump tinged buffy at top***. Sandy-brown or grey-buff female often lacks black throat, recalling female Black-eared, but separated by rump and tail pattern. Often confiding. **Voice** Calls include a soft whistle.

Black-eared Wheatear *Oenanthe hispanica* PM

L 13.5–15. Common spring migrant, uncommon in autumn. Spring, mainly Mar–early May, peaks early–mid Apr, occasionally to mid Jun. Breeding reported, but not substantiated. In autumn, occasionally from Aug, mainly Sep–Oct, peaking mid Sep. Black-throated individuals outnumber pale-throated forms. **Habitat** Sparsely vegetated or stony slopes; headlands and foothills. **ID** *O. h. melanoleuca* ('Eastern' Black-eared Wheatear) occurs in Cyprus. In flight, by ***black on sides of tail***; black tail-band sometimes of uneven width. Small, structure much as Cyprus Wheatear. Male of black-throated form (frequent in region) can be confused with male Finsch's Wheatear but ***black of throat/ear-coverts does not join black of wings and shoulder***, mantle whitish (summer) or buffish-grey (autumn). Female has whitish chin, usually with slight rustiness below and ***sandier mantle***. From first-winter Northern Wheatear by tail pattern. Some males do not wear white by spring and appear similar to western race (*hispanica*), making them hard to identify with certainty. **Voice** Rather variable: dry, scratchy twittering mixed with clear whistles, often heard in flight. Calls include a hard *tack*, a soft sneeze and a characteristic buzzing (like an angry fly!).

Northern Wheatear
♂ 1st-winter
♂
♀
♂
Isabelline Wheatear
1st-winter
ad
♀
♂
Desert Wheatear
1st-winter
♂
♀
♂
♂
♀
Black-eared Wheatear

Cyprus Wheatear *Oenanthe cypriaca* MB

L 13.5. Very common and fairly widespread breeder. Cyprus breeding endemic (formerly treated as race of Pied Wheatear). Occurs Mar–Nov, occasionally in winter. Mainly late Mar–mid Oct, widely on territory from Apr at all elevations. Family parties common by late May. Reportedly hosts Common Cuckoo. Departs mid Sep–Oct, some remaining to mid Nov. **Habitat** Headlands on migration. Favours hilly locations and old village houses at all elevations to breed; very arboreal; high density in montane forests. Nests among rocks, or in holes in pipes, electric metre boxes, machinery crevices, open sheds, under eaves or any suitable walls/roadside banks, trees and cliffs. **ID** Spring male has black mantle and wings meeting black face and bib, with whitish crown (closely resembling male Pied Wheatear) but warmer buff below black 'bib'. ***Best identified by song, narrower white rump, broader black tip to tail and short primary projection***. Breeding female has slaty to grey-brown upperparts and dark crown (all white in male), surrounded by whitish stripe; in worn plumage appears 'black and white', resembling male at a distance. Autumn adult has grey-buff fringes to black mantle and throat, dark crown, buff-white supercilium and ***deep rusty-buff underparts***. First-winter similar but has more, and broader, pale fringes. Juvenile (mid May–mid Sep) has dark grey-brown upperparts, head (without eyestripe) and upper breast, with cream spotting; mid-breast to undertail greyish-white with dark fringes; scalloping on breast and upper flanks; white rump and black tail as adult. **Voice** Song ***recalls*** **cicada** but less harsh, a lengthy purring *bizz-bizz-bizz*, often ending in high-pitched, drawn-out piping note, delivered from high song post (usually a treetop but also overhead wires, rooftops, pylons, aerials, higher ground, etc.), exceptionally in flight. Calls include a hard *tack*.

Pied Wheatear *Oenanthe pleschanka* V

L 14–16. Vagrant. **ID** Slender, often perches on bushes. ***From Mourning Wheatear by absence of white panel in open wing***. Breeding male very similar to Cyprus Wheatear (which see). Rare white-throated form, '*vittata*', closest to 'Eastern' Black-eared Wheatear but mantle black. Autumn male has black back and throat fringed buffish, dark crown with buff-white supercilium, and buff underparts. Female like female Black-eared Wheatear, but ***upperparts duller, cooler brown-grey***, some show large dark greyish 'bib' (absent in female Black-eared); autumn female also by dark brown breast-sides merging with greyish 'bib'. First-winter female with difficulty from Black-eared by colder upperparts; crown, mantle and shoulders ***scalloped with rows of pale fringes*** (mainly absent in Black-eared). First-winter female from Northern Wheatear by ***black extending up sides of tail;*** black tail-band ***sometimes of uneven width***. **Voice** Calls hard *tack*, dry *trrrlt* or dry, sneezed *snerr*.

Mourning Wheatear *Oenanthe lugens* V

L 13.5. Vagrant. **ID** Race *lugens* has occurred. Sexes very similar; resembles adult male Cyprus Wheatear when perched, but ***undertail-coverts apricot (all ages); in flight prominent whitish wing panel***. Compared to Cyprus Wheatear, black bib is smaller, underparts pure white (warm-buff in Cyprus), primary-coverts narrowly tipped white and black band on tail lacks black extensions on outer feathers of Pied. In autumn lacks pronounced pale fringes to black throat and mantle. Juvenile lacks black mantle and throat, has blackish ear-coverts, a broad buffish-white bar on greater coverts and tips of otherwise black primary-coverts. **Voice** Call *check-check*; alarm *peet-peet*.

Kurdish Wheatear *Oenanthe xanthoprymna* V

L 15. Vagrant. Mainly Nov–mid Apr. **Habitat** Headlands and peninsulas; boulder-strewn hillsides and rocky outcrops. **ID** Male with whitish supercilium, black throat, sides of head and neck, merging with blackish-brown wing-coverts, ***rufous rump with white sides to tail***; black band at tip of tail narrow; and ***little white in wings***. Female lacks black throat. Usually solitary; bounding hops and slight downward tail flicks; flies with tail closed. **Voice** Calls include *steu-steu-steu*, alarm a short dry *zuk* or *zvee-tuk*. [Alt: Rufous-tailed Wheatear.]

♂
♀
♂
♂ 1st-winter
ad autumn
Cyprus Wheatear
♂
juv
♂
♂ 1st-winter
'vittata'
♀
♂
Pied Wheatear
ad
♂
lugens
Mourning Wheatear
♀ dark-throated
Kurdish Wheatear

Hooded Wheatear *Oenanthe monacha* pm

L 17. Rare spring migrant. Occurs late Mar–mid May, occasionally other months. **Habitat** Headlands, sea cliffs and ruins, rarely far from coast. **ID** *Large, slender, long-tailed and long-winged wheatear with long bill and buoyant, almost butterfly-like flight,* recalling Spotted Flycatcher when catching prey, sometimes in long sallies. Adult male distinctive, with *whitish crown, black below to centre of breast* and, except black central feathers, *nearly all-white tail with just black corners.* Autumn and first-winter male have creamy-buff crown, whitish fringes to black throat, wing-coverts and mantle, and lower underparts, rump and sides of tail tinged buffish. Female sandy brownish-grey above *merging into cream-buff rump, tail-coverts and sides of tail,* in which central feathers and corners are dark brown; whitish-grey underparts washed buff on breast-sides, flanks and undertail-coverts. In autumn and first-winter, female rump and underparts can appear reddish-buff with almost reddish-brown tail-sides, but absence of dark terminal tail-band separates from Red-tailed Wheatear. **Voice** Female utters a *whit-whit* or repeated *jiirp* like a fledgling Eurasian Blackbird; also a *wit-awheat-wheet-wheet* or *whee-whee-whee-wheeoo.*

White-crowned Wheatear *Oenanthe leucopyga* v

L 17. Vagrant. Records Feb–May. **Habitat** Likely on coastal cliffs, gorges and headlands. **ID** Large wheatear with sexes similar; glossy *black with black underparts down to legs, and, in many adults, a white crown*; young and some adults have black crown, but size, long bill and *white tail-sides with black corners diagnostic* (no black terminal band). Some black-crowned birds show a few white feather tips, eventually developing a white crown. In male Hooded Wheatear, which has similar tail pattern, black extends only to centre of breast. **Voice** Variable song of whistling tuneful notes, sometimes scratchy, often with imitations of other locally occurring species; common phrase *viet-viet-dreeit-deit*, slightly descending, but much variation. Call *peeh-peeh.*

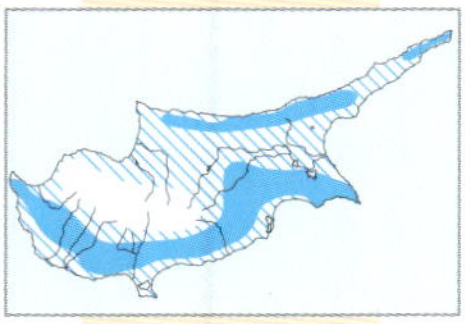

Finsch's Wheatear *Oenanthe finschii* WV, pm

L 15. Fairly common winter visitor and scarce passage migrant. Occurs late Sep–Apr, mainly late Oct–mid Mar and mostly males. **Habitat** Hillsides with scattered boulders and stone walls; favours grazed areas, sometimes near abandoned villages. Occasional on passage at headlands. **ID** Heavier and stockier than Cyprus and Black-eared Wheatears; male from former by *narrow buff-white stripe on mantle and back to white rump*; from latter by *larger black 'bib' broadly connected to black wings* (beware Black-eared with head 'sunk' between shoulders); from both by *terminal tail-band of even width* (no black extension on sides) and *pale greyish flight feathers below, appearing translucent above.* Some females have variable blackish throat (sometimes only lower throat). *Flight feathers and primary-coverts often finely pale-tipped* (both sexes). Female often shyer than male. Ground-dwelling, perches infrequently in trees. Frequently *bows low, cocking tail, repeatedly spreading and lowering it slowly.* **Voice** Alarm *tack*; also *che-che-che.*

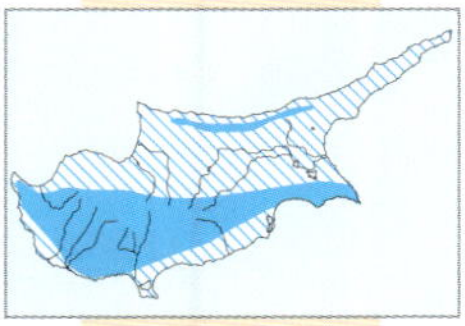

Dunnock *Prunella modularis* wv, pm?

L 14.5. Uncommon winter visitor and probable passage migrant. Occurs late Oct–Apr, peaking Dec–Jan. Scarce and shy; overlooked in montane forests. **Habitat** Thorny scrub and woodland undergrowth, on low or high ground. **ID** *Warm brown with distinctly streaked upperparts, sombre grey head and breast*; often has spotted wingbar. Juvenile has brown-spotted crown and ear-coverts, with most of underparts streaked (attains adult plumage by early autumn). Often in trios on ground; wings or tail flicked jerkily. **Voice** Call short, discordant, strident *tjee.* In flight, utters high ringing *dididididr.* [Alt: Hedge Sparrow.]

♂ winter
♂
♀
♀
Hooded Wheatear
♂
ad
juv
ad
imm
White-crowned Wheatea
♀
♀
♂
Finsch's Wheatear
ad
♂
Dunnock

PLATE 84: SPARROWS

House Sparrow *Passer domesticus* RB, PM? wv?

L 15. Abundant breeding resident. Locally dispersive and possibly migratory. Noticeable movements Mar–early May and Sep–Nov, often with Spanish Sparrow flocks. Nests Jan–Jun and in autumn. **Habitat** Mainly towns, villages, farmland or around isolated buildings at all elevations; often away from habitation. Nests colonially in holes in walls, buildings and trees. **ID** Widespread, familiar. ***Male has grey crown, chestnut-brown sides of head, greyish cheeks and variable black 'bib'***; wing-coverts rufous-chestnut with broad white bar on median coverts. Female and juvenile buffish-brown above, ***boldly streaked darker on mantle***, pale greyish below and have creamy supercilium. Female hard to separate from female Spanish Sparrow, which see. Often in flocks. **Voice** Call and song monotonous cheeping or chirping notes of varying pitch; alarm a scolding *cherr-r-r-r-r-r*.

Spanish Sparrow *Passer hispaniolensis* RB, PM, WV

L 15. Very common breeding resident, passage migrant and winter visitor. Passage mid Mar–Apr; largest numbers Oct (when flocks move along south coast). Nests Feb–Jun. **Habitat** Villages and rural areas; open farmland in winter; also lower pine forests. Nests in base of larger birds' nests (e.g. corvids), hole in wall or in palm tree, often colonially. **ID** Resembles House Sparrow (with which it hybridises); larger bill evident. ***Male has rufous-brown crown (pale fringed and duller in winter) and large black 'bib' extending as bold black streaks on breast and flanks (all pale-fringed in winter); back boldly streaked black, merging with black of breast***. Female not safely separated from House Sparrow, but supercilium and underparts whiter, with breast and flanks grey-streaked. Gathers in large, compact flocks on passage and in winter. **Voice** Song recalls House Sparrow's, but faster and more piercing, especially at colonies and roost.

Eurasian Tree Sparrow *Passer montanus* v

L 14. Vagrant. Most records late Aug–early May. **Habitat** Peninsulas and headlands; sometimes inland or with other sparrow species. **ID** Smaller, more delicate than House Sparrow with ***chestnut crown, black spot on whiter cheeks and smaller black 'bib'; two white wingbars*** and grey-brown rump. **Voice** Song a series of repeated *tweet* notes. Calls clearer and harder than House Sparrow's, higher pitched; flight call a hard *tek-tek-tek*; also *tchu-wit*, *pilp* and a hard *chik*.

Dead Sea Sparrow *Passer moabiticus* FB, v

L 12. Former breeder (last bred 1994), now a vagrant. **Habitat** Reedbeds, tamarisks and eucalyptus. **ID** ***Small; distinctive head pattern with dark grey cheeks and crown contrasting with long white and rusty supercilium, and pale yellow moustachial bordering small black bib. Female resembles small House Sparrow*** with slightly shorter wingtip, but undertail-coverts dark-spotted. **Voice** Song rolling *tri-rirp, tri-rirp* or *tlir-tlir-tlir*; calls include a short *chilp* like House Sparrow, a harder, metallic, high-pitched *tlip* or *trrirp* and occasional buzzing *tzzeeer*.

Pale Rockfinch *Carpospiza brachydactyla* v

L 15. Vagrant. Occurs Mar–May, mainly late Apr. **Habitat** Headlands; occasionally river valleys near coast. **ID** ***Unstreaked, grey-brown, lark-like sparrow with whitish wing panel and wingbars, long primary projection, short dark tail with white tip*** (obvious in flight), strong pale bill with curved culmen, and dark eye. In flight, the long, triangular wings give a lark-like appearance. Feeds in manner of lark, otherwise hops almost upright. **Voice** Flight call a soft trill, recalling distant European Bee-eater; also *piyee* or *twee-ou*. [Alt: Pale Rock Sparrow, Hill Sparrow.]

Rock Sparrow *Petronia petronia* v, wv?

L 15. Vagrant or rare winter visitor; Nov–Apr, mainly mid Dec–mid Mar. **Habitat** Montane slopes, stony ground with scattered bushes; also dry riverbeds and farmland. Vacates mountains of Turkey and Iran in winter. **ID** Thickset with large head, short tail, longer primary projection and thick legs. ***Boldly striped head with pale crown-stripe and conspicuous long buffy-white supercilium; strongly streaked breast and flanks, and white-tipped tail diagnostic. Yellowish spot on upper breast hard to see***. Juvenile lacks yellow spot. Flight undulating; often in loose flocks. Active on ground, running among rocks. **Voice** Characteristic metallic *twe-uh-ee*; whispering nasal chatter from flocks in flight. Song a hoarse *vee-veip*.

♂ winter
♀
♂
House Sparrow
♀
♂ hybrid
Spanish x House
♂
♂ winter
Spanish Sparrow
♂ winter
♀
♂
Dead Sea Sparrow
ad
Eurasian Tree Sparrow
ad
Pale Rockfinch
♂
♀
Rock Sparrow

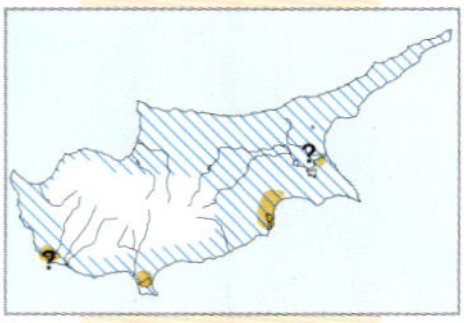

Western Yellow Wagtail *Motacilla flava* PM, mb

L 16.5. Common passage migrant and uncommon breeder (*feldegg*). Occurs in many races and forms. On Cyprus, the commonest subspecies are *feldegg* (Black-headed Wagtail), *flava* (Blue-headed Wagtail) and *thunbergi* (Grey-headed Wagtail), with *feldegg* breeding in variable numbers. **Habitat** Damp meadows, ploughed and irrigated fields, farmland and marshes, usually at low altitudes. **ID** In spring and summer, head colour of males varies as described below; hybrids occur, making some identifications uncertain. Females all similar: ***yellowish below with whitish throat***, and most are unidentifiable to race. Many birds in autumn are simply too difficult to identify, especially females. **Voice** Calls: *flava* high *see-u*; *thunbergi* slightly sharper; *beema* buzzier.

Black-headed Wagtail *M. f. feldegg* Very common passage migrant and scarce breeder. Occurs mid Feb–Oct; spring passage mainly Mar–early May, peaking late Mar–early Apr; less common in autumn Aug–Oct, peaking early Sep. Occasional in winter. The commonest race in spring. Breeds Mar–Jun. **Habitat** Breeds in wet salt scrub. **ID** Male has head and nape glossy black, sharply demarcated from dark mossy-green mantle, entire underparts bright yellow; duller in winter. Female superficially like male, often almost monochrome, being dark-headed with greyish upperparts and yellow-sullied underparts. **Voice** Song a loud, simple *sree-srriep*, with stress on second syllable. Call a strong ringing *psreee-u*. **Note** Includes '*melanogrisea*' (Eastern Black-headed Wagtail), see below.

Blue-headed Wagtail *M. f. flava* Very common passage migrant (and may breed). Spring mainly mid Mar–May, peaking early–mid Apr; autumn Aug–Oct, peaking Sep. **ID** Blue-grey crown and ear-coverts; long white supercilium; sides of throat sometimes white.

Grey-headed Wagtail *M. f. thunbergi* Fairly common passage migrant. Spring mainly late Mar–mid May (5–10 days later than *feldegg*), peaking early Apr; autumn late Aug–Sep, peaking mid Sep. **ID** Dark grey crown, not sharply demarcated from upperparts but contrasting with darker ear-coverts, usually no supercilium or white on chin and throat.

Sykes's Wagtail *M. f. beema* Uncommon passage migrant. Spring mid Mar–May, peaking May; rare in autumn, and at all times much confused with colour variants of Blue-headed Wagtail. **ID** Pale greyish crown and ear-coverts, long broad white supercilium and whitish chin. Some first-winter birds lack yellow below (rear of underparts normally tinged yellow in those of other races), are greyish above (normally tinged olive-brown in other races) and have fairly prominent whitish wingbars, thus resembling first-winter Citrine Wagtail (call also rather similar).

Eastern Black-headed (White-chinned) Wagtail *M. f. 'melanogrisea'* Rare passage migrant. **ID** Resembles *feldegg* but back paler olive, underparts paler yellow, chin and sides of throat white.

Ashy-headed Wagtail *M. f. cinereocapilla* Rare migrant (and has bred). **ID** Recalls *thunbergi* but throat whitish. **Voice** Harsh, ringing *psreee-u*.

Yellow-headed Wagtail *M. f. lutea* Vagrant. **ID** Entire head yellow, some with olive crown and ear-coverts, upperparts yellow-olive.

Yellow-crowned (British) Yellow Wagtail *M. f. flavissima* Vagrant. **ID** Recalls *lutea*, but always shows olive-green crown and back.

Hybrids

'*superciliaris*' (*feldegg* × *flava* or *beema*) Fairly common passage migrant. Usually with groups of Black-headed and Blue-headed Wagtails. **ID** Dark slaty-blue head with clean white supercilium (of variable size and intensity), submoustachial stripe and broad yellowish wingbars.

'*dombrowskii*' (*feldegg* × *flava* or *beema*) Fairly common passage migrant. Occurs with groups of Black-headed and Blue-headed Wagtails. **ID** Similar to *flava*, but blackish ear-coverts.

'*xanthophrys*' (*feldegg* × *lutea*) Rare passage migrant. **ID** Supercilium yellow, crown, lores and ear-coverts contrastingly and solidly dark, underparts yellow. (Not illustrated.)

Western Yellow Wagtail
juv/1st-winter
flava
♂
♀
flavissima
♂
'melanogrisea'
♂
cinereocapilla
♂
thunbergi
♂
feldegg
♂
lutea
♂
♀
feldegg
beema
♂
'dombrowskii'
♂
'superciliaris'
♂

Citrine Wagtail *Motacilla citreola* pm

L 18. Uncommon passage migrant. Spring passage Feb–Jun, mainly mid Mar–May, peaking Apr. Mostly young in autumn, late Jul–mid Nov, peaking late Aug–early Sep. **Habitat** Fresh and brackish wetlands, marshy edges, freshly cut fields, sewage and waste-water pools. **ID** Male unmistakable; *M. c. werae* (from Turkey) predominates, which has no (or reduced) neck-band and less rich yellow underparts; ***broad pure white double wingbars characteristic at all ages***. *M. c. citreola* has ***bright yellow head*** and underparts, ***black neck-band and grey upperparts***. Female has ***greyish upperparts*** (tinged olive-brown in Yellow Wagtail), ***yellow supercilium surrounding grey-brown cheeks merging on sides of neck into yellowish throat***. First-winter has grey upperparts, ***pale surround to ear-coverts, often pale forehead and lores, whitish underparts without yellow***, prominent wingbars and broadly pale-fringed dark tertials. **Voice** Call pronounced ***buzzy tsreep***; sometimes a double *zielip*; some eastern '*flavas*' rather similar.

Grey Wagtail *Motacilla cinerea* PM, WV, rb

L 18. Common passage migrant, winter visitor and scarce breeding resident. Common on autumn passage mid Aug–early Dec, peaking Oct–early Nov; fairly widespread in winter Dec–Mar; spring passage imperceptible mid Mar–early May. Breeds Apr–Jul. **Habitat** Migrants occur by rivers and streams, freshwater ponds (even swimming pools), canals, rubbish dumps and sewage plants. Breeds by streams in montane forests and wooded upland valleys, where some remain all year. Nests in hole in bank or wall. **ID** In all plumages has ***yellow vent and undertail-coverts, greyish back, bold white translucent bar at base of flight feathers (in flight), very long tail and extremely undulating flight***. Male in summer ***has black throat and white submoustachial stripe***; female has less black or even whitish throat; first-year has white throat and buff, not yellow, breast. ***Legs pinkish-brown*** (black in other *Motacilla*). Tail-wagging more pronounced than in other *Motacilla*. **Voice** Call distinctive, resembles White Wagtail's but harder, more metallic and high-pitched, ***a piercing tzi-lit or tsiziss, or loud chink***, usually given in bounding flight. Song a distinctive series of sharp mechanical notes.

White Wagtail *Motacilla alba* PM, WV, ob

L 18. Very common passage migrant, winter visitor and has bred. Occurs mid Sep–May. Autumn migration late Sep–mid Dec, peaking Oct. Winter visitors depart by late Feb; spring passage mid Feb–mid May, peaking Mar. Occasional in summer. **Habitat** Damp areas, mainly on low ground, including sewage plants, meadows, farmland, fields and cultivation; also urban areas and large gardens. Roosts on roofs and trees in towns and cities. **ID** Easily identified by ***grey, black and white plumage***, with grey back contrasting with black nape (male) or crown (female), white face and ear-coverts contrasting sharply with black throat and breast. Juvenile, and especially winter birds, have whitish underparts with ***prominent black half-moon collar on breast***. Flanks show some grey, concolorous with or paler than mantle. Flight undulating, when often calls. Several vagrant races have occurred (see below) plus *ocularis* ('East Siberian Wagtail'; not illustrated). **Voice** Disyllabic call slightly metallic, a hard *tse-lit* with stress on second syllable. Juvenile's call more metallic, often with more syllables. Song comprises call notes, a lively garbled twittering.

Pied Wagtail *Motacilla alba yarrellii* v

L 18. Vagrant. **Habitat** As White Wagtail. **ID** Similar to White Wagtail, but male in summer has ***jet-black upperparts*** (often scruffier in winter); female and first-winter male sooty-grey above with ***dark rump. Flanks often mainly dark grey*** (often extensively so), unlike similar White Wagtail. **Voice** As White Wagtail.

Masked Wagtail *Motacilla alba personata* v

L 18. Vagrant, mainly spring and autumn. **Habitat** Open areas, often near habitation, cultivation and especially near cattle. **ID** *M. a. personata* has ***black head and breast merging on sides of neck and ear-coverts, leaving smaller white area around face*** (same pattern in both sexes, but darker in male); white fringes to wing-coverts, tertials and secondaries much broader than in White Wagtail. **Voice** Call more metallic than White Wagtail, approaching Grey Wagtail.

winter
♂ summer
Citrine Wagtail
1st-winter
♀
Grey Wagtail
♂ summer
1st-winter
♂
♀
White Wagtail
♂ summer
Pied Wagtail
♂ winter
ad
juv
Masked Wagtail

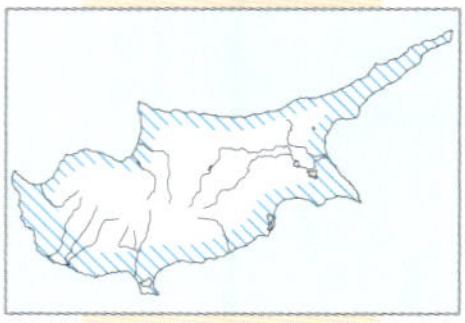

Richard's Pipit *Anthus richardi* pm

L 18. Uncommon passage migrant. Occurs Mar–early May, peaking mid Apr; and Sep–early Dec, peaking early Oct. Occasional in winter. **Habitat** Damp meadows, short grass or recently cut fields; also marshy grassland. **ID** Large, robust pipit; adult from smaller, sandier Tawny Pipit by *longer tail and legs, stouter bill, typically more upright stance, prominently dark-streaked brown upperparts, long hindclaw, streaked breast and characteristic call*. From first-winter Tawny Pipit, some still with well-streaked breast, by *pale lores* (dark in Tawny, bleaching whitish). More undulating flight and longer tail produce more wagtail-like flight than Tawny; *frequently hovers just before landing* (Tawny rarely does). Rather shy and flighty. **Voice** Flight call a loud, harsh throaty *schreeip*; also less distinctive Tawny-like *tjiirrup*.

Blyth's Pipit *Anthus godlewskii* v

L 17. Vagrant. **ID** *Very like Richard's Pipit but slightly smaller, shorter tailed (noticeable in flight) and shorter legged*, with slightly shorter, deep-based but pointed bill (Richard's has culmen decurved at tip); supercilium often shorter. Outer tail feathers also differ (see plate). Crown neatly streaked, lores pale. Mantle and breast streaked, latter often with a neat gorget; flanks sometimes buffy. *Hindclaw medium length (long in Richard's). Median coverts of adult (if not worn) and first-winter with some renewed to adult-type (but not otherwise) have dark centres sharper, more squarely cut-off from pale tips than Richard's, producing a prominent pale bar* (in Richard's, dark centres protrude centrally, with pale fringes more diffuse); first-winter from young Tawny Pipit by pale lores, more streaked upperparts, warmer browner plumage, distinctly streaked breast, with *peachy-toned* flanks and underparts, and narrower, clearer-cut pale tertial fringes. Drops into grass without hovering. **Voice** *High-pitched shreuu or spzeeoo, unlike Richard's flight call, closer to slightly hoarse sweeuu call of flava wagtails; also a short chep, chip, chup or pip reminiscent of Tawny Pipit; sometimes joined into unique schreuu-chup chup*.

Tawny Pipit *Anthus campestris* PM

L 16.5. Common passage migrant. Spring passage mid Feb–May, peaking late Mar–Apr; fewer in autumn late Aug–Nov, peaking mid Sep–mid Oct. Occasional in winter. **Habitat** Favours dry areas, mostly fallow fields and marginal land; also foothills in west and south. **ID** Slim, largish, upright pipit with *relatively long tail, legs and bill. Adult has poorly streaked sandy upperparts, nearly unstreaked breast*, plain sandy wings with conspicuously dark-centred median coverts, and bold whitish supercilium. First-winter, still with streaked breast, *lacks bold flanks streaks of most smaller pipits*. Legs pinky-salmon. Flight undulating; runs quickly, stopping suddenly. **Voice** Typical flight call *sparrow-like chilp or chirrup*. Song, usually in undulating flight, simple, thin and metallic, *zriiliu, zseer-lee* or *ziu-ziirliu*.

Long-billed Pipit *Anthus similis* v

L 17. Vagrant. **Habitat** Most likely at headlands in spring. **ID** Large; from Tawny Pipit by *creamier-buff flanks and vent*, less pronounced malar and *creamy-buff outer tail feathers* (which bleach whitish to resemble Tawny). *Legs pinky-yellow*. Bill longer with more drooping tip than Tawny, and at least culmen dark. *Upperparts grey-brown, tinged olive in some*, pale edges and dark centres to wing-coverts and tertials generally less pronounced than Tawny, supercilium often narrower and ear-coverts plainer brown. *Long-, broad- and dark-tailed* in flight; *when perched tail often flicked upwards and fanned outwards*. Seldom well-streaked on breast and upperparts. From Richard's by plainer head pattern, less upright stance, *less streaked plumage and buff outer tail feathers* (always white in Richard's). Confiding. **Voice** Flight call a loud, clipped *tjuip* or *che-vee*; also rich *tchup* and quiet, soft *tchut*. Rising and falling song given in undulating flight, *duiit-diuuu, peet-trueet* or *shreep chew-ee*.

ad
1st-winter
Richard's Pipit
1st-winter
Blyth's Pipit
juv
ad
Tawny Pipit
ad
1st-winter
Long-billed Pipit

PLATE 88: SMALLER PIPITS

Meadow Pipit *Anthus pratensis* WV, PM

L 14.5. Very common winter visitor and passage migrant. Occurs late Sep–Apr. Autumn passage late Oct–early Dec, declining Mar, few to late Apr. Imperceptible spring passage. **Habitat** Irrigated fields, crops and grassy meadows; coasts and foothills. **ID** More slightly built than Tree Pipit, with ***long hindclaw***, orange-brown legs; from non-breeding Red-throated Pipit by ***unstreaked rump, broken breast and flanks streaks, call*** and less contrasting mantle pattern. Flies with undulations and changes of direction; on ground flicks tail nervously. Regularly perches. **Voice** Flight call thin, nervous, *sit–sii–sit* or *tsis–sip*.

Tree Pipit *Anthus trivialis* PM

L 15. Common passage migrant. Occurs Mar–May (occasionally earlier), peaking Apr; fewer, mid Aug–mid Nov, peaking late Sep–mid Oct. **Habitat** Usually dry ground with scattered trees; coasts and foothills. Not dependent on grassy meadows and crops. **ID** Stockier than Meadow Pipit, with shorter tail and deeper-based bill. Less olive above, buff on breast with whiter belly; breast boldly streaked, flanks finely so; dark malar and creamy bold submoustachial, fainter moustachial. ***Short, pale hindclaw; legs pinkish-salmon***. Bounding flight; on ground pumps tail, unlike Meadow's nervous flicks. **Voice** Flight call distinctive, a short *bzeez* or *speez*; alarm a repeated *stit*. Contact note distinctive, a quiet *tip*.

Olive-backed Pipit *Anthus hodgsoni* V

L 14.5. Vagrant. **Habitat** Dry ground, mainly capes and headlands. **ID** Resembles Tree Pipit, with broad supercilium, white behind eye, buff in front, edged black above, pronounced small white and black spots on rear ear-coverts; upperparts greenish, vaguely or diffusely streaked unless worn; breast boldly streaked black; bold blackish flanks streaks. Tail-wagging usually more pronounced than Tree Pipit. **Voice** Flight call a thin *tzeez*, usually drawn-out, slightly stronger at start.

Red-throated Pipit *Anthus cervinus* PM, wv

L 15. Common passage migrant, scarce in winter. Occurs Sep–early Jun. Spring passage Mar–May, mainly Apr–early May; autumn Sep–early Dec, peaking mid–late Oct. **Habitat** Irrigated crops, cut fields and seasonal wetlands; flooded salt scrub, rubbish dumps, dams and water treatment plants. **ID** Adult has variable amount of ***pinkish or reddish-buff on face, supercilium, throat and upper breast***, retained in autumn but rarely in winter; female can have beige throat. From Meadow Pipit in winter by ***call; boldly streaked rump***, creamy or white breast and flanks, ***flanks with 2–3 bold, unbroken stripes; malar ends in dark triangle on side of throat***; upperparts ***bolder blackish and creamy-white stripes***. **Voice** Distinctive and diagnostic, high-pitched, drawn-out *pseeee*, slowly dying away.

Buff-bellied Pipit *Anthus rubescens* V

L 15.5. Vagrant. **Habitat** Grassy fields, near water. **ID** *A. r. japonicus* (from Siberia) has occurred; resembles dark winter Water Pipit; ***upperparts dark olive or grey-brown, faintly streaked; breast and flanks usually rather boldly streaked; legs pale brown or reddish-brown; white eye-ring obvious, malar streak often broader, ending in black patch on sides of neck***. In spring, underparts pinkish-buff, lightly spotted, ***flanks strongly streaked***; upperparts unstreaked olive-grey. **Voice** High-pitched, short *tripp*, similar to Meadow Pipit.

Water Pipit *Anthus spinoletta* PM, WV

L 16. Common passage migrant, less common in winter. Occurs late Sep–mid May. Autumn passage mid Oct–early Dec, peaking late Oct–mid Nov. Spring passage Mar–mid May, peaking mid–late Mar. **Habitat** Marshes, farms, rubbish dumps, dams, pools and salt lakes. **ID** *A. s. coutellii* ('Caucasian Water Pipit') occurs. Largish with ***dark brown (appearing black) legs*** and indistinctly streaked greyish mantle. In winter browner above, with ***whitish underparts sparsely streaked, almost unstreaked in some***. In summer ***breast tinged rosy-pink and practically unstreaked***; creamy supercilium pronounced. **Voice** Call a sharp ringing *tsrieh* or *bzisp*.

ad
Meadow Pipit
ad
Tree Pipit
ad
Olive-backed Pipit
ad summer
ad summer
japonicus
ad summer
ad winter
1st-winter
Buff-bellied Pipit
Red-throated Pipit
ad summer
coutellii
ad winter
Water Pipit

White-winged Snowfinch *Montifringilla nivalis* v

L 17.5. Vagrant. **ID** Superficially resembles Snow Bunting *Plectrophenax nivalis* (not recorded in Cyprus). All races have *small black bib in summer. In flight shows much white in wing and tail but primaries and central tail black*; white less extensive in female and juvenile. Bill black in summer male, yellowish at other seasons and ages. Perches upright, often jerking tail. **Voice** Song in flight or when perched, a jerky, repeated *sitticher-sitticher*; flight call a nasal harsh *tsweek*; also a purring *prrt*.

Eurasian Crimson-winged Finch *Rhodopechys sanguineus* v

L 16. Vagrant. **ID** From Trumpeter Finch by larger size, *brown mottled upperparts, breast and flanks, black crown and yellow bill* (greyish-horn in non-breeder). Male has pink rump and face (pink on face lacking in winter) and *large rosy-pink flash in wings* striking in combination with white underwing-coverts in flight. Female duller, has paler wing panel, lacks pink in face and has dull yellow bill in breeding season. Flight over distance deeply undulating. **Voice** Flight call a trisyllabic *tchu-che-ly* or soft *chee-rup - tureep - tureep*, and Woodlark-like *dy-lit-di-lyt*. Song a melodious trill from ground or in flight.

Hawfinch *Coccothraustes coccothraustes* wv

L 18. Uncommon winter visitor. Variable numbers, with noteworthy irruptions some years. Occurs from late Oct (occasionally earlier), mainly mid Nov–Mar, with stragglers to early or mid Apr. **Habitat** Dense and light mixed woodland. In some winters reliably in the Troodos, or at lower elevations in well-wooded villages or riverine valleys with mature trees. Scarce east of Limassol. **ID** Rather large, short-tailed, heavy finch readily identified by *large, powerful steel-grey bill* (horn-coloured in winter), rufous and brown plumage with *black on face and chin*, and *broad white wingbar*. Female duller with less rufous-orange head and mainly grey (not black) flight feathers. Juvenile lacks black on face, and is rather greyish with scalloped underparts. *In fast, bounding flight has white translucent wingbar and white-tipped short tail*. Loose groups perch high on exposed branches, and feed on ground with other finches. Can be rather shy. **Voice** Song a simple series of sharp notes on different pitch. Calls in flight an explosive *pzik* and a strange, thin *fzeeik*.

Red Crossbill *Loxia curvirostra* RB, wv?

L 17. *L. c. guillemardi* ('Cyprus Crossbill') is a common resident. Rare influxes reported, very occasionally at lower altitudes away from main forests, Sep–Dec, perhaps involving migrants. Breeding depends on food availability, but at least Jan–Jun, often earlier and occasionally at lower elevation (once near Nicosia). **Habitat** Generally confined to highest areas of the Troodos, above 1,100m, where it feeds on seeds of *Pinus nigra*. Visits puddles to drink, otherwise invariably in canopy or outer branches of conifers, dropping spent cones. Nest usually high, but few details. **ID** Large finch with *short tail, stout head and heavy bill crossed at tip*. Adult male blotchy red with strong yellow or orange tinge to underparts, with brown wings and tail; *in flight, strikingly red rump. Female greenish* with greyish-brown wings and tail, and slightly streaked underparts; *in flight, greenish-yellow rump*. Juvenile similar to female but distinctly streaked throughout. First-year gradually attains adult plumage. In flight, clumsy and large-headed with short forked tail; *often in loud flocks*, but can be very elusive. **Voice** Typical flight call a loud, hard *kip-kip*; often silent when feeding but may give similar but softer, subdued *chip-chip* notes. Song irregular with mechanical twitters, trills and admixed calls developing into *cheeree-cheeree-choop-chip-chip-chip-cheeree*. [Alt: Common Crossbill.]

winter
♂
♂
White-winged Snowfinch
♀
ad
♂
Hawfinch
Eurasian Crimson-winged Finch
♂
♀
juv
♂
guillemardi
Red Crossbill
nominate

Common Chaffinch *Fringilla coelebs* RB, PM, WV

L 15. Common resident, passage migrant and winter visitor. Winter visitors (mostly females) present late Oct–mid or late Mar, stragglers to mid Apr; passage Feb–early Mar. Breeds Mar–Jun. **Habitat** Resident in montane forests; breeds locally in lowland woods, notably on Akrotiri Peninsula; migrants mainly on low ground and foothills, in olive groves, orchards, vineyards and open country with bushes and trees. **ID** Male has ***blue-grey head, chestnut back, pinkish breast and double white wingbars***. Resembles Brambling but no orange in plumage, has prominent white wingbars and ***lacks white rump***. Nondescript female from all other finches by white wingbars, white outer tail feathers and unstreaked brownish-olive upperparts. Undulating flight with wings closed every few beats. Largely feeds on ground with rather jerky shuffling movements. Forms flocks on passage and in winter. **Voice** Song a short, accelerating trill, ending in a flourish. Calls vary considerably but typically a loud metallic *fink*, repeated loud *huit* and, in flight, a soft *jup jup*.

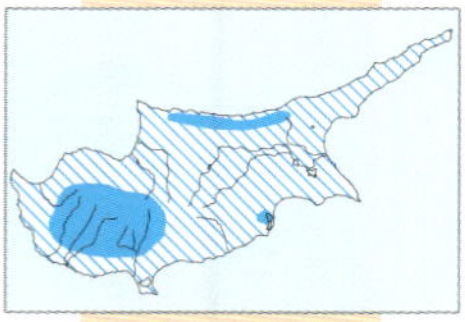

Brambling *Fringilla montifringilla* wv, pm

L 15. Scarce winter visitor and passage migrant in varying numbers; almost absent some years. Occurs late Oct–Mar, occasionally to early May. **Habitat** Wooded areas at all elevations; regularly in montane forests, especially the Troodos. **ID** Resembles Common Chaffinch but has ***rusty-orange breast, shoulder and wingbar, white belly, mottled upperparts and, in flight, white rump; pale patch on nape and yellowish bill with dark tip***. Mixes freely with Chaffinches and buntings in winter foraging areas. Flight as Chaffinch but more erratic. **Voice** Distinctive, in flight or perched, a loud, unmusical nasal *jehp*, or softer repeated *tip* or *pip*.

European Greenfinch *Carduelis chloris* RB, PM, WV

L 14.5. Very common resident, passage migrant and winter visitor. First bred in 1960s and still increasing. Autumn passage erratic, some from late Aug, mainly mid Sep–mid Nov; spring passage mainly late Jan–early Apr. Breeds Feb–Jul. **Habitat** Conifer trees; carob and almond groves, orchards, forest edges and vineyards; weedy coastal scrub, ploughed and stubble fields. Nests in conifers, usually cypress. **ID** Stocky, fairly large-headed finch with stout, conical pale bill. ***In all plumages by bright yellow edges to primaries and outer tail feathers, very obvious in flight***. Male olive-green with yellow-green underparts and rump; female duller grey-green, and juvenile olive grey-brown and streaked. Relatively short tail in undulating flight. Gregarious, feeds on ground in loose groups. **Voice** Song, from treetop or in bat-like flight, a powerful canary-like trill; sometimes a prolonged descending nasal *tweeesh*. Call a short *tjup* or *teu*, alarm a hoarse *de-u-wee*.

Desert Finch *Rhodospiza obsoleta* v

L 14.5. Vagrant. **ID** Pale, greyish-buff with striking wing pattern and short black bill. Flight feathers have pronounced white fringes; pink panel on coverts and base of secondaries. Male has black lores and base of bill. Female drab, less pink, white and black in wing, and lacks face patch. In flight, from below, looks pale with diffuse translucent wing panel; above, white primary patch with pink on secondaries and greater coverts. Feeds on ground. **Voice** Call a quiet, soft, melodic *prrrrrl*, suggesting distant bee-eater.

Eurasian Bullfinch *Pyrrhula pyrrhula* v

L 14. Vagrant. **Habitat** Woodland, particularly conifers. **ID** Male has ***pinkish-red underparts, grey back, black cap, broad white wingbar and stout black bill***. Female much duller, buffish below and thus less gaudy than male. ***In flight, white rump and all-black tail***. Generally stays in cover or thickets, but may feed in open on buds in spring. **Voice** A low, soft and quiet, slightly falling, piped whistle, *dwoo*, easily imitated.

♂
♀
♂
Common Chaffinch
♂
♀ winter
♂ winter
Brambling
juv
♀
European Greenfinch
♂
♀
Desert Finch
♀
♂
♀
Eurasian Bullfinch
♂

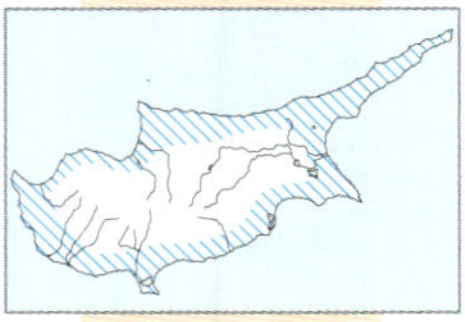

Trumpeter Finch *Bucanetes githagineus* pm

L 14. Rare spring passage migrant. Less than annual. Most records Mar–May, peaking mid–late Apr; occasional in autumn. **Habitat** Mostly near coasts, particularly on rocky headlands. **ID** Small, ground-dwelling finch with large head, stout bill and rather short tail. Male distinctive by ***grey head, pinkish wash on forehead, underparts, rump and wings, and orange-red bill***. Non-breeding male, female and juvenile/first-winter plain sandy grey-buff with slightly paler rump, blackish-grey wings and tail, with paler feather edgings and pale yellowish-brown bill; some breeding females develop faint pink wash; ***legs always orangey-flesh***. **Voice** Song distinctive, a drawn-out nasal, wheezing *cheeeee*; call a short *chee* or *chit*; in flight a soft *weechp*, most calls with buzzing quality.

Common Rosefinch *Carpodacus erythrinus* v

L 14. Vagrant. **Habitat** Anywhere on passage; often at wetlands with trees and tall scrub. **ID** Compact finch with ***stout bullfinch-like bill*** and round head. Adult male easily identified by ***red head, breast and rump contrasting with brown upperparts***. Female, first-summer male (often sings) and juvenile dull olive-brown, lightly streaked above and ***more heavily streaked below, with two whitish or buffish wingbars***, but no panel in flight feathers; uniform head with conspicuous dark beady eye. Often sings from prominent position, otherwise inconspicuous. **Voice** Song diagnostic, a clear, lively loud whistle: *vii'-dji'vii-di-djiv-viuuu* (rendered 'pleased to meet youuu').

Common Linnet *Linaria cannabina* RB, WV, PM

L 13. Common resident, winter visitor and passage migrant. Surges of migrants occur throughout winter and spring, notably late Oct–early Nov and mid Jan–mid Mar; most depart by Apr. Otherwise rather patchy, and absent from many areas. Breeds late Jan–May, occasionally in autumn. **Habitat** Stony, rough and ploughed land from coastal plains to hills, favouring thistle seeds; also weedy, sandy foreshores. **ID** ***In male, greyish head, chestnut back and crimson-red forehead and breast characteristic***. Female and juvenile lack red and chestnut, are warm brown above with dark streaks, visible also on breast and flanks (especially in juvenile, which also lacks grey on head). ***At all ages has white-fringed primaries and outer tail feathers, which, together with dark grey-brown bill, streaked central throat, and pale area around eye and on cheeks, separate it from Twite***. Flight irregularly undulating. In pairs and flocks in winter. **Voice** Song a varied musical twitter, interspersed by whistling nasal notes, uttered from exposed perch. In flight, a nasal, sharp *tett* or *tetteret*, often followed by short trill or whistle, *trreu* or *tru-kee-wou*. Anxiety call a soft mellow *di-ge-dee*.

Common Redpoll *Acanthis flammea* v

L 13. Vagrant. **ID** Small, compact, greyish-brown finch with heavily streaked body and fairly long, forked tail. Distinctive features include ***black bib, short yellowish bill and red forecrown*** (sometimes lacking). Male has pink breast and may be slightly pink on rump. Female and first-winter darker, browner with heavier streaked flanks. See Twite. Very active when foraging on ground or in trees; agile, often hangs upside-down on thin branches. **Voice** Flight call a rapidly repeated, hard *djä-djä-djä*; when perched a soft rising *dyuee*.

♀
♂
Common Rosefinch
♀
Trumpeter Finch
♂
1st-winter
♀
♂
♂ winter
Common Linnet
Common Redpoll

European Goldfinch *Carduelis carduelis* RB, WV, PM

L 14. Very common resident, winter visitor and passage migrant. Autumn migration mid Sep–Nov, peaking late Oct–early Nov; variable numbers in winter, with main spring passage mid Feb–early Apr. Breeds Jan–Jun; sometimes multi-brooded to Sep. **Habitat** Open country, meadows, cultivation and forests. Migrants mix with local residents in winter, foraging in flocks on thistle and other seed heads. Breeds at all altitudes, especially in pine forests. **ID** Colourful; adult ***readily identified by bold red, black and white head pattern, broad yellow bar on black wings, particularly obvious in flight, when white rump also conspicuous***. Juvenile has pale greyish body and unmarked head; identified by wingbar, white rump and call. Often in small flocks outside breeding season. A popular cage bird. **Voice** Call, often given in flight, a characteristic, repeated, liquid *tick-le-lit* with stress on last note; rasping notes sometimes heard from flocks. Song a liquid twitter, mixed with calls.

Red-fronted Serin *Serinus pusillus* V

L 12. Vagrant. **Habitat** Meadows or rocky hillsides; often with groups of European Serin. **ID** Small, stubby, ***dark-streaked finch with sooty-brown head and breast, and almost luminous orange patch on forehead***; in bounding flight shows dark fore-body with rather pale underparts and underwing. Juvenile has pale orange-buff face, and upperparts and underparts heavily streaked black. **Voice** Call a European Serin-like ripple, but softer, a drawn-out *drrrrt*.

European Serin *Serinus serinus* PM, WV, RB

L 12. Very common passage migrant, winter visitor and locally common resident. Autumn passage patchy, with waves of migrants late Aug–early Dec; spring passage late Jan–early May, peaking mid Feb–early Mar, often in large flocks. Breeds late Feb–Jul. **Habitat** Migrants occur in almond and carob groves, vineyards and conifers at all elevations; flocks in irrigated fields and meadows. Breeds in coniferous woodland, mainly in uplands, but increasingly in small stands of pine and cypress at low elevations. **ID** Small yellowish, streaked finch with stubby bill; ***yellow rump, supercilium, face and breast, white belly with bold flanks streaks and all-dark tail***. Female much duller, barely showing any yellow; generally streaked olive-grey with washed-out yellowish-buff face and, like male, yellow rump; black flanks streaks on whitish underparts; double wingbar can be fairly obvious or nearly absent. First-winter as adult. Often in small parties outside breeding season. Flight fast and undulating, almost skipping. **Voice** Song jingling, a fast twitter like the sound of shaking keys, from exposed perch or in bat-like song flight. Flight call distinctive, a metallic tinkled *zr-r-litt*.

Eurasian Siskin *Spinus spinus* WV

L 12. Fairly common winter visitor, late Oct–early Apr, mostly late Dec–Mar. In varying numbers, with large influxes some years. **Habitat** Upland coniferous forests, increasingly at low elevations; associated with Oriental alder and plane trees in riverine forests; also wooded villages with cypress and other evergreens. **ID** Small, finely built finch; ***male yellowish-green with black forehead and chin, and broad yellow wingbar, sides of tail and rump, particularly obvious in flight***. Female (which lacks black on head) is olive-brown above, white below but has tinge of yellow-green, which, together with yellowish in wings, tail and on rump, aid identification. Back and flanks streaked dark at all ages. Often feeds upside-down in cone-bearing trees. Appears small and short-tailed in undulating flight, often in flocks, individuals erratically changing position and direction. **Voice** Call a drawn-out, ***high-pitched tsee-u***; from feeding flocks a fast, dry *kettkett* and twittering notes. Song, sometimes in butterfly-like flight, a variable fast twitter ending in a drawn-out wheezing note, often includes mimicry.

juv
ad
European Goldfinch
ad
1st-summer
1st-winter
ad
Red-fronted Serin
♀
juv
♂
♂
European Serin
♀
♂
Eurasian Siskin

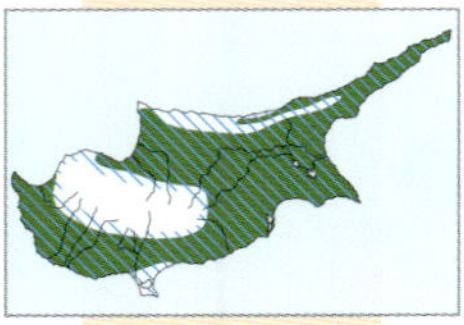

Corn Bunting *Emberiza calandra* RB, PM, WV

L 18. Common breeding resident, passage migrant and winter visitor. Complicated movements involve winter visitors overlapping resident and possible migrant breeders. In autumn, influx from late Oct, with numbers increasing through winter; spring passage Mar–mid Apr. Breeds Jan–Jun, **Habitat** Meadows, cereal fields and vineyards, from sea level to foothills, where it also breeds. Present all year in hilly farmland; may congregate around dams and reservoirs in summer. **ID** Bulky with fairly large head, *large conical bill and no white in tail*. Underparts heavily streaked, often forming irregular black spot on breast; often, pale submoustachial stripe and variable ill-defined malar. *Heavy, rather fluttering flight, legs often dangling*; lacks white on trailing edge of wings of Eurasian Skylark, with which it can be confused in flight. Often perches on telegraph wires or bushes; flocks in non-breeding season. **Voice** Song, often from exposed perch, accelerates and ends with monotonous, almost insect-like sound, *tick-tick-tzek-zee-zri-zizizizi*. Call a hard, almost clicking *twik* or *tritt*, sometimes quickly strung together in series.

Yellowhammer *Emberiza citrinella* WV

L 16.5. Uncommon to rare winter visitor, Nov–Apr, mostly Jan–Mar. **Habitat** Mainly in higher pine forests; occasionally lightly wooded areas at lower elevations. **ID** Medium-sized bunting with grey bill and *chestnut rump at all ages. Male's yellow crown and throat diagnostic; female's yellowish moustachial, belly, sometimes supercilium and central crown separate from similar female Pine Bunting*. First-winter (sometimes with difficulty) from Pine Bunting by yellow on belly or vent and on primary edges; a minority can lack any yellow below. Hybridises with Pine Bunting in Siberia; hybrid males often (but not always) show white of head sullied yellow. Separated from first-winter Cirl Bunting by combination of some yellow in streaked plumage and chestnut rump (grey-brown in Cirl) or call. Forms flocks in winter. **Voice** Metallic, sharp *staeup*, and fast *twe-tic* in flight.

Pine Bunting *Emberiza leucocephalos* V

L 16.5. Vagrant. Records Nov–Mar. **Habitat** As Yellowhammer. **ID** Identical to Yellowhammer in structure and size. *Male's black-bordered white crown and cheeks in otherwise chestnut head and throat diagnostic* (subdued in winter). Grey-brown female has streaked crown, mantle, breast and flanks; chestnut rump like Yellowhammer; *some show a little whitish on crown, and chestnut on whitish throat. Many females, and all first-winters, resemble Yellowhammer, but yellow replaced by white, including on belly and fringes to primaries*; lesser coverts more uniform grey-brown and bill more frequently bicoloured than Yellowhammer (dark grey upper, pale grey lower mandible). Some males from hybrid zone (Siberia) often have yellow tinge to plumage (head, fore-body and primaries), others appear mainly white-headed; female Yellowhammer/Pine Bunting hybrids probably defy separation. Often associates with Yellowhammers in winter. **Voice** Some calls identical to Yellowhammer; also utters a nervous *trr-rrr-rrr-ick*.

Rock Bunting *Emberiza cia* V, pm? wv?

L 16. Vagrant or rare passage migrant and winter visitor. Altitudinal migrant (probably from Turkey). Occurs Nov–Apr. **Habitat** Apparently regular on higher slopes of Pentadaktylos Mountains, in sparse woodland with well-vegetated shrub layer and limestone rocky outcrops; anywhere on passage. **ID** Male distinctive; *black-striped pale grey head, unstreaked throat, boldly black-streaked mantle, grey bill, white sides to tail and orange belly (at all ages)*. Female more diffusely patterned but similar to male, with head-stripes brownish. Juvenile recalls young Yellowhammer (both have chestnut-brown rump). Unobtrusive, often flushes from ground only when close. **Voice** Call a thin *zip* or *zeet*.

Corn Bunting
♀
hybrid Pine Bunting
x Yellowhammer
♂
♀ 1st-winter
Yellowhammer
♀
♂ spring
♂ winter
Pine Bunting
♀
♂
1st-winter
Rock Bunting

PLATE 94: BUNTINGS II

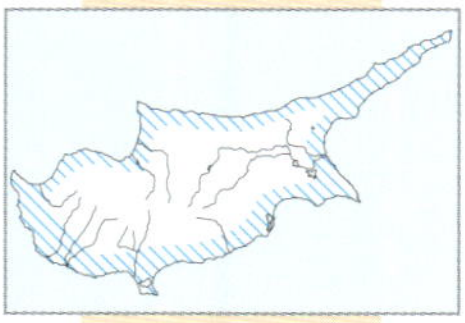

Cinereous Bunting *Emberiza cineracea* pm

L 16.5. Scarce spring passage migrant, infrequent in autumn. Occurs mid Mar–Apr, peaking late Mar–early Apr; occasionally late Aug–mid Sep. **Habitat** Grassy slopes with rocky outcrops; also headlands. **ID** *E. c. cineracea* ('Western Cinereous Bunting') occurs. Medium-sized with ***blue-grey bill and narrow yellow-white eye-ring. Male has unstreaked yellowish-green crown, sides of face and throat, otherwise greyish, including unstreaked rump***. Olive grey-brown female streaked below, as winter male; ***throat buffy-white; streaked crown tinged grey-brown***. Browner, more streaked first-winter (with streaked grey-brown rump) from other grey-brown buntings by lack of rufous rump, pale eye-ring, less patterned cheeks and ear-coverts, and absence of chestnut-brown scapulars. **Voice** Call short, metallic *kjip* or *djib*. **Note** *Western Cinereous breeds only in western Turkey and predominates on passage through Near East*.

Ortolan Bunting *Emberiza hortulana* PM

L 16.5. Common passage migrant. In spring mid Mar–early Jun, mainly late Mar–early May, peaking mid Apr (when often in large flocks); less common mid Aug–Oct, peaking early Sep, when juvenile and autumn plumages confusable with Cretzschmar's Bunting. **Habitat** Low ground and foothills with scattered bushes and trees; in autumn also at dams and pools. **ID** Medium-sized with ***white eye-ring, pink bill, boldly streaked mantle and blackish-brown tertials with clear-cut pale chestnut*** (adult) ***or creamy-buff notched edges. Male from similar Cretzschmar's Bunting by olive-grey head and breast-band framing pale yellowish throat***. Female duller and more streaked; from female Cretzschmar's by pale yellow-buff throat, brownish (not rufous-tinged) rump and pale buff vent. First-winter more streaked above and on breast than female; hard to separate from similar Cretzschmar's, but colour of rump and vent useful. Migrates in flocks. **Voice** Call a hollow, soft *plet* or *büb* and a slightly falling *sliie*.

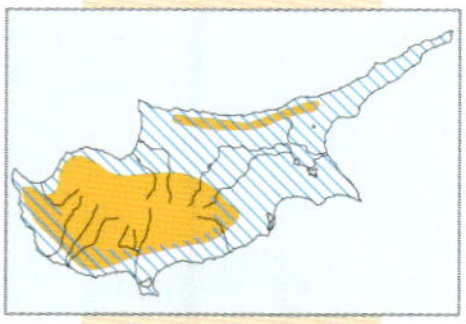

Cretzschmar's Bunting *Emberiza caesia* MB, PM

L 16.5. Common breeding visitor and passage migrant. Spring passage Mar–Apr, peaking late Mar; autumn passage Aug–early Nov, mainly late Aug–mid Sep; difficult to separate migrants from local dispersal. Breeds mid Mar–Jul. **Habitat** Migrants occur in open areas with scattered bushes and trees, on low ground and foothills; often at headlands. Breeds in scrub with larger bushes; also boulder-covered lightly wooded hillsides and open pine forest, usually above 150m. **ID** Resembles bright Ortolan Bunting. ***Male has bright blue-grey head and breast-band framing rusty-orange throat; rump rufous, vent and undertail-coverts rufous-buff***, blackish-brown tertials have clear-cut rufous notched fringe. Female duller, often with traces of grey on head and streaked breast; ***from female Ortolan by rusty throat*** (sometimes yellowish-buff) and warmer rump and vent. First-winter resembles Ortolan but ***yellow-buff throat often tinged rufous (adult plumage gained quickly)***. Colour of rump and vent useful to separate first-winter females. **Voice** Short song recalls Ortolan, but thinner, more metallic, the second part confined to a single *diu-diu-diu-duu*. Calls Ortolan-like, but sharper, *blep*; metallic *tlik* or *tlev*.

Yellow-breasted Bunting *Emberiza aureola* V

L 15. Vagrant. **ID** Medium-sized with pale, stout, conical bill and white in tail-sides. ***Male by black face and upper throat, dark chestnut-brown upperparts and narrow breast-band on otherwise bright yellow underparts, also prominent white wingbars***. Female and first-winter ***by prominent creamy supercilium, faint pale central forecrown-stripe with dark brown lateral crown-stripes*** (less defined in first-winter), ***strongly dark-bordered pale cheeks and fairly distinct white wingbars***. Rump grey-brown, heavily streaked. Underparts yellowish-white or at least sullied buffy (mainly first-winter) with whitish undertail-coverts; flanks and sometimes breast faintly streaked. Female sometimes has warm buffish-yellow neck-sides; upperparts streaked blackish and greyish-yellow (sometimes tinged rusty), often with creamy braces. **Voice** Two main calls heard on passage, a short *tik*, not unlike Little Bunting, and a softer, metallic *tsiu* or *tsip*.

1st-winter
♂
Cinereous Bunting
cineracea
♂ 1st-summer
♀
1st-winter
♂
Ortolan Bunting
♂ 1st-winter
♂
Cretzschmar's Bunting
♀
♀
juv
♂
Yellow-breasted Bunting
♂ 1st-summer

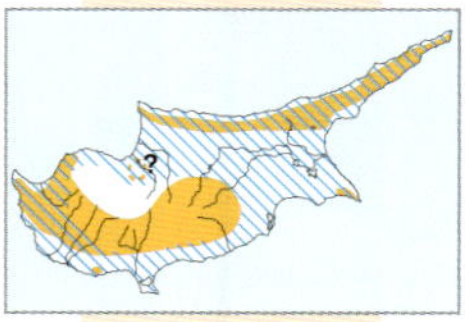

Black-headed Bunting *Emberiza melanocephala* MB, PM

L 16.5. Fairly common breeding visitor and passage migrant. Occurs late Mar–early Oct, mainly mid Apr–early Jul; some passage late Apr–early May. Usually arrives in east of island, spreading west from mid Apr. Scarce autumn passage Sep–early Oct. Breeds late Apr–Jun, most gone by early Aug. **Habitat** Vineyards, citrus groves and cereal fields, where it also breeds; common in Akamas and Makheras forests. **ID** Heavy-bodied, stout-billed bunting with *no white in tail. Male has black head, yellow underparts and unstreaked chestnut upperparts*; black of head subdued by pale fringes in autumn. Female and first-winter lack black, bright chestnut and yellow; *upperparts usually warm-toned olive-grey* and underparts generally tinged yellow, especially undertail-coverts. Sometimes in loose groups on spring passage. Females shy and retiring. **Voice** Variable song recalls Common Whitethroat's in staccato form, but strong and typically bunting-like, starting with short harsh notes, ending with more ringing metallic notes: *tsi-tia-tia-tia-terlu-terlu-terlu*. Call a sparrow- or Tawny Pipit-like *tjilp*; also a metallic *tlev*.

Common Reed Bunting *Emberiza schoeniclus* WV, pm

L 15.5. Uncommon winter visitor and scarce passage migrant. Numbers vary annually. Occurs Oct–mid Apr, mainly Nov–Feb. **Habitat** Almost exclusively reedbeds, of any size and usually at low elevations. **ID** Breeding male has *black hood and bib, white half-collar and white moustachial*. Female has *broad black malar and whitish supercilium* on otherwise brown head. First-winter resembles female but cheeks more mottled, breast and flanks more streaked. From Little Bunting by *brownish (not chestnut) ear-coverts, convex culmen, inconspicuous eye-ring, chestnut lesser coverts and by call*. When perched, jerks and spreads tail nervously. In flight, undulations short and irregular, compared to more steady flight of most buntings. Some Reed Buntings breeding in Turkey and Iran (*E. s. pyrrhuloides*) are *more strongly marked above with noticeably stouter bill*. **Voice** Calls include a fine, drawn-out *tsii-u* and a metallic, slightly voiced *bzü*; in flight a low nasal *bäh*.

Little Bunting *Emberiza pusilla* v

L 13. Vagrant. **Habitat** On passage, damp bushy and grassy areas. **ID** Small and secretive. *Adult by reddish-chestnut crown, supercilium and ear-coverts, framed by black lateral crown-stripe, and ear-coverts surrounding narrow whitish eye-ring; white underparts with narrow black streaks on breast*, whitish wingbars usually obvious. First-winter from Common Reed Bunting by *uniform pale rusty cheeks, bill shape (straight culmen), whitish eye-ring, black malar barely reaching bill, grey-brown lesser coverts, better-defined blacker streaks on whiter underparts and call*. From Rustic Bunting by *absence of rufous streaks below and grey-brown rump* (rusty in Rustic). **Voice** Call a single, hard, metallic *tik*, like Rustic Bunting but slightly sharper.

Rustic Bunting *Emberiza rustica* v

L 14.5. Vagrant. **Habitat** On passage, generally in damp, bushy or wooded areas. **ID** Fairly robust bunting, streaked rufous below and has longish bill with straight culmen. *Male unmistakable, with black head broken by bold white supercilium and nape spot, bright rusty nape merging into neck-sides with rusty red breast-band; flanks extensively streaked rusty; rump rusty-red*. Female has less blackish head with paler median crown-stripe and ear-coverts. First-winter confusable with Common Reed Bunting but *has rufous streaks on breast and flanks, rufous rump, different bill colour and shape, bolder supercilium behind eye, whiter wingbars* and different call. **Voice** Call a short, distinct *zib* or *zik*, less sharp than Little Bunting.

♀
1st-winter
♀
♂ winter
♂
Black-headed Bunting
♀
Common Reed Bunting
♂ winter
♂
1st-winter
1st-winter
♂ spring
Little Bunting
♂ 1st-winter
♀ 1st-winter
♂
Rustic Bunting

REFERENCES AND FURTHER READING

BirdLife Cyprus publishes a Cyprus Bird Report annually and a monthly newsletter entitled *The Monthly Checklist*, which includes papers and detailed systematic lists.

The Ornithological Society of the Middle East, the Caucasus and Central Asia, OSME (osme.org) publishes *Sandgrouse* twice yearly. It contains topical articles and latest sightings and is usually where new birds for the region are published.

Bannerman, D.A. & Bannerman, W.M. 1958. *Birds of Cyprus*. Oliver & Boyd, Edinburgh & London.

Beton, D., Bates, C., Easterbrook, M., Pegler, N. & Snape, R. 2016. *2012 Checklist*. North Cyprus Society for Protection of Birds and Nature (KUŞKOR), Cyprus.

Blair, M., Preddy, S., Al-Sirhan Alenezi, A., Porter, R.F. & Aspinall, S.J. 2019. Ornithological Society of the Middle East, the Caucasus and Central Asia (OSME) Regional List (ORL) v4.4. osme.org/ORL

Chandler, R. 2009. *Shorebirds of the Northern Hemisphere*. Christopher Helm, London.

Charalambides, M. 1995. Cyprus Warbler *Sylvia melanothorax*. P. 437 in Tucker, G.M. & Heath, M.F (eds.) *Birds in Europe, Their Conservation Status*. BirdLife International, Cambridge, UK.

Clement, P. & Rose, C. 2015. *Robins and Chats*. Christopher Helm, London.

Dawes, J. & Whaley, D. 2012. Finsch's Wheatear - discovery of previously unreported wintering sites. *The Monthly Checklist* January 2012: 16–19.

Flint, P.R. 2019. Long-term changes in the numbers and abundance of regularly breeding land bird species on Cyprus: a review. *Sandgrouse* 41(1): 36–70.

Flint, P.R. 2000. The Hobby *Falco subbuteo* as a breeding summer visitor to Cyprus. *North Cyprus Bird Report 1999*: 41–43.

Flint, P.R. & McArthur, A. 2014. Is the Sardinian Warbler *Sylvia melanocephala* displacing the endemic Cyprus Warbler *S. melanothorax* on Cyprus? *Sandgrouse* 36(1): 63–109.

Flint, P.R. & Richardson, C. 2017. A review of the residential status of the endemic Cyprus Scops Owl *Otus cyprius*. *Sandgrouse* 39(2): 177–181.

Flint, P.R. & Stewart, P.F. 1983. *The Birds of Cyprus: An Annotated Check-list*. BOU Checklist No. 6. British Ornithologists' Union, Tring (second, revised, edn. published 1992).

Flint, P.R., Whaley, D., Kirwan, G.M., Charalambides, M., Schweizer, M. & Wink, M. 2015. Reprising the taxonomy of Cyprus Scops Owl *Otus (scops) cyprius*, a neglected island endemic. *Zootaxa* 4040(3): 301–316.

Forsman, D. 2016. *Flight Identification of Raptors of Europe, North Africa and the Middle East*. Christopher Helm, London.

Gill, F. & Donsker, D. 2018. IOC World Bird List, Version 8.2. worldbirdnames.org

Hellicar, M.A., Anastasi, V., Beton, D. & Snape, R. 2014. *Important Bird Areas of Cyprus*. BirdLife Cyprus, Nicosia.

Ieronymidou, C., Collar, N.J. & Dolman, P.M. 2012. Endemic Cyprus Warbler *Sylvia melanothorax* and colonizing Sardinian Warbler *Sylvia melanocephala* show different habitat associations. *Ibis* 154: 248–259.

Kassinis, N., Gücel, S., Charalambidou, I., Turkseven, N., Fuller, W., Kuyucu, A. & Yorgancı, H. 2010. *Waterbirds in Cyprus 2008/2009*. UES-CCEIA/BA/CGF, Nicosia.

Kephalas, A. 2011. *Birds in June (Reports and Numbers)*. Privately published, Frenaros.

Kirwan, G.M., Boyla, K., Castell, P., Demirci, B., Özen, M., Welch, H. & Marlow, T. 2008. *The Birds of Turkey.* Christopher Helm, London.

Kourtellarides, L. 1998. *Breeding Birds of Cyprus.* Bank of Cyprus, Nicosia.

Message, S. 2005. *Waders of Europe, Asia and North America.* Christopher Helm, London.

Mitchell, D. 2017. *Birds of Europe, North Africa and the Middle East, An Annotated Checklist.* Lynx Edicions, Barcelona.

Olsen, K.M. 2018. *Gulls of the World: A Photographic Guide.* Christopher Helm, London.

Olsen, K.M. & Larsson, H. 2003. *Gulls of Europe, Asia & North America.* Christopher Helm, London.

Papazoglou C., Crabtree, A., Panayides, P., Ioulianou, F. & Conway, G. 2017. Tracking Cyprus Wheatears *Oenanthe cypriaca* with geolocators. *Sandgrouse* 39(2): 130–137.

Porter, R. & Aspinall, S. 2010. *Birds of the Middle East.* Second edn. Christopher Helm, London.

Richardson, C. (ed.) 2005–12. *Cyprus Bird Reports 2004–2011.* BirdLife Cyprus, Nicosia.

Richardson, C. 2014. The status, frequency and abundance of passerine passage migrants and seasonal visitors in Cyprus 2003–2013. *Sandgrouse* 36(2): 197–236.

Roth, T. 2008. Outward (autumn) bird migration at the Southeastern Peninsula and Cape Greco, Cyprus: the phenologies of regular migrants. *Sandgrouse* 30(1): 77–89.

Shirihai, H. 1996. *The Birds of Israel.* Academic Press, London.

Shirihai, H. & Svensson, L. 2018. *Handbook of Western Palearctic Birds.* Vols. 1–2. Christopher Helm, London.

Sparrow, D.J. & John, E. (eds.) 2016. *An Introduction to the Wildlife of Cyprus.* Terra Cypria, Limassol.

Stylianou, J. (ed.) 2013–17. *Cyprus Bird Reports 2012–2016.* BirdLife Cyprus, Nicosia.

Stylianou, J. (ed.) 2018. *Birds of Cyprus Checklist 2018.* BirdLife Cyprus, Nicosia.

Svensson, L., Mullarney, K. & Zetterström, D. 2009. *Collins Bird Guide.* Second edn. HarperCollins, London.

Whaley, D.J. & Dawes, J.C. 2003. *Cyprus Breeding Birds Atlas.* Privately published, Cyprus.

CHECKLIST OF THE BIRDS OF CYPRUS

❑	Greylag Goose	*Anser anser*	wv
❑	Greater White-fronted Goose	*Anser albifrons*	wv
❑	Lesser White-fronted Goose	*Anser erythropus*	v
❑	Red-breasted Goose	*Branta ruficollis*	v
❑	Mute Swan	*Cygnus olor*	wv
❑	Bewick's Swan	*Cygnus columbianus bewickii*	v
❑	Whooper Swan	*Cygnus cygnus*	v
❑	Egyptian Goose	*Alopochen aegyptiaca*	v
❑	Common Shelduck	*Tadorna tadorna*	WV, pm
❑	Ruddy Shelduck	*Tadorna ferruginea*	wv
❑	Gadwall	*Mareca strepera*	pm, wv, ob?
❑	Eurasian Wigeon	*Mareca penelope*	wv, pm
❑	Mallard	*Anas platyrhynchos*	RB, PM, WV
❑	Northern Shoveler	*Spatula clypeata*	WV, PM, ob
❑	Northern Pintail	*Anas acuta*	WV, PM
❑	Garganey	*Spatula querquedula*	PM, ob
❑	Eurasian Teal	*Anas crecca*	WV, PM
❑	Marbled Duck	*Marmaronetta angustirostris*	v, FB
❑	Red-crested Pochard	*Netta rufina*	wv, pm, ob
❑	Common Pochard	*Aythya ferina*	WV, pm
❑	Ferruginous Duck	*Aythya nyroca*	WV, pm, rb
❑	Tufted Duck	*Aythya fuligula*	wv, pm
❑	Greater Scaup	*Aythya marila*	v
❑	Common Scoter	*Melanitta nigra*	v
❑	Common Goldeneye	*Bucephala clangula*	v
❑	Smew	*Mergellus albellus*	v
❑	Goosander	*Mergus merganser*	v
❑	Red-breasted Merganser	*Mergus serrator*	wv
❑	Ruddy Duck	*Oxyura jamaicensis*	v? E/I?
❑	White-headed Duck	*Oxyura leucocephala*	wv, pm?
❑	Chukar Partridge	*Alectoris chukar*	RB, E/I
❑	Black Francolin	*Francolinus francolinus*	RB
❑	Common Quail	*Coturnix coturnix*	PM, wv, mb
❑	Black-throated Diver	*Gavia arctica*	v
❑	European Storm Petrel	*Hydrobates pelagicus*	v
❑	Scopoli's Shearwater	*Calonectris diomedea*	PM
❑	Yelkouan Shearwater	*Puffinus yelkouan*	PM

❑	Little Grebe	*Tachybaptus ruficollis*	WV, PM, RB
❑	Red-necked Grebe	*Podiceps grisegena*	v
❑	Great Crested Grebe	*Podiceps cristatus*	wv, pm, ob
❑	Horned Grebe	*Podiceps auritus*	v
❑	Black-necked Grebe	*Podiceps nigricollis*	WV, PM ob
❑	Greater Flamingo	*Phoenicopterus roseus*	PM, WV ob?
❑	Lesser Flamingo	*Phoeniconaias minor*	v
❑	Black Stork	*Ciconia nigra*	pm
❑	White Stork	*Ciconia ciconia*	pm
❑	Glossy Ibis	*Plegadis falcinellus*	PM, mb
❑	Eurasian Spoonbill	*Platalea leucorodia*	pm
❑	Eurasian Bittern	*Botaurus stellaris*	pm, wv
❑	Little Bittern	*Ixobrychus minutus*	PM, ob
❑	Black-crowned Night Heron	*Nycticorax nycticorax*	PM, wv, mb
❑	Striated Heron	*Butorides striata*	v
❑	Squacco Heron	*Ardeola ralloides*	PM, mb
❑	Western Cattle Egret	*Bubulcus ibis*	RB, PM, WV
❑	Grey Heron	*Ardea cinerea*	PM, WV, ob?
❑	Purple Heron	*Ardea purpurea*	PM
❑	Great Egret	*Ardea alba*	pm, wv
❑	Little Egret	*Egretta garzetta*	PM, wv, mb
❑	Great White Pelican	*Pelecanus onocrotalus*	pm
❑	Dalmatian Pelican	*Pelecanus crispus*	v
❑	Northern Gannet	*Morus bassanus*	pm
❑	Pygmy Cormorant	*Microcarbo pygmaeus*	v
❑	Great Cormorant	*Phalacrocorax carbo*	PM, WV
❑	European Shag	*Phalacrocorax aristotelis*	RB
❑	Western Osprey	*Pandion haliaetus*	PM
❑	Black-winged Kite	*Elanus caeruleus*	v
❑	Lammergeier	*Gypaetus barbatus*	v
❑	Egyptian Vulture	*Neophron percnopterus*	pm
❑	European Honey Buzzard	*Pernis apivorus*	PM
❑	Crested Honey Buzzard	*Pernis ptilorhynchus*	v
❑	Griffon Vulture	*Gyps fulvus*	rb, pm?
❑	Cinereous Vulture	*Aegypius monachus*	v, FB
❑	Short-toed Snake Eagle	*Circaetus gallicus*	pm
❑	Bateleur	*Terathopius ecaudatus*	v
❑	Lesser Spotted Eagle	*Clanga pomarina*	pm
❑	Greater Spotted Eagle	*Clanga clanga*	v

❑	Booted Eagle	*Hieraaetus pennatus*	pm
❑	Eastern Imperial Eagle	*Aquila heliaca*	v, FB
❑	Golden Eagle	*Aquila chrysaetos*	v
❑	Bonelli's Eagle	*Aquila fasciata*	rb, pm?
❑	Levant Sparrowhawk	*Accipiter brevipes*	pm
❑	Eurasian Sparrowhawk	*Accipiter nisus*	PM, WV, ob
❑	Northern Goshawk	*Accipiter gentilis*	rb, pm
❑	Western Marsh Harrier	*Circus aeruginosus*	PM, WV
❑	Hen Harrier	*Circus cyaneus*	WV, pm
❑	Pallid Harrier	*Circus macrourus*	PM
❑	Montagu's Harrier	*Circus pygargus*	PM
❑	Red Kite	*Milvus milvus*	v
❑	Black Kite	*Milvus migrans*	PM, wv
❑	White-tailed Eagle	*Haliaeetus albicilla*	v
❑	Rough-legged Buzzard	*Buteo lagopus*	v
❑	Long-legged Buzzard	*Buteo rufinus*	RB, pm, wv?
❑	Common Buzzard	*Buteo buteo buteo*	WV, pm
❑	– Steppe Buzzard	*Buteo buteo vulpinus*	PM
❑	Great Bustard	*Otis tarda*	v
❑	MacQueen's Bustard	*Chlamydotis macqueenii*	v
❑	Little Bustard	*Tetrax tetrax*	v
❑	Water Rail	*Rallus aquaticus*	WV, PM, ob
❑	Corn Crake	*Crex crex*	pm
❑	Little Crake	*Porzana parva*	PM, wv
❑	Baillon's Crake	*Porzana pusilla*	pm, ob
❑	Spotted Crake	*Porzana porzana*	PM
❑	Grey-headed Swamphen	*Porphyrio poliocephalus*	v
❑	Allen's Gallinule	*Porphyrula alleni*	v
❑	Common Moorhen	*Gallinula chloropus*	PM, WV, RB
❑	Eurasian Coot	*Fulica atra*	WV, PM, RB
❑	Demoiselle Crane	*Grus virgo*	PM
❑	Common Crane	*Grus grus*	PM, wv
❑	Eurasian Stone-curlew	*Burhinus oedicnemus*	RB, PM, wv?
❑	Eurasian Oystercatcher	*Haematopus ostralegus*	pm
❑	Black-winged Stilt	*Himantopus himantopus*	PM, MB, wv
❑	Pied Avocet	*Recurvirostra avosetta*	pm, wv
❑	Northern Lapwing	*Vanellus vanellus*	WV, PM
❑	Spur-winged Lapwing	*Vanellus spinosus*	RB, PM
❑	Sociable Lapwing	*Vanellus gregarius*	v

❑	White-tailed Lapwing	*Vanellus leucurus*	v
❑	European Golden Plover	*Pluvialis apricaria*	WV, PM
❑	Pacific Golden Plover	*Pluvialis fulva*	v
❑	Grey Plover	*Pluvialis squatarola*	wv, pm
❑	Common Ringed Plover	*Charadrius hiaticula*	PM, wv
❑	Little Ringed Plover	*Charadrius dubius*	PM, MB, wv
❑	Kittlitz's Plover	*Charadrius pecuarius*	v
❑	Kentish Plover	*Charadrius alexandrinus*	RB, PM, WV
❑	Greater Sand Plover	*Charadrius leschenaultii*	pm, wv
❑	Caspian Plover	*Charadrius asiaticus*	pm
❑	Eurasian Dotterel	*Charadrius morinellus*	pm
❑	Eurasian Woodcock	*Scolopax rusticola*	wv
❑	Jack Snipe	*Lymnocryptes minimus*	pm, wv
❑	Great Snipe	*Gallinago media*	pm
❑	Common Snipe	*Gallinago gallinago*	PM, WV
❑	Black-tailed Godwit	*Limosa limosa*	PM, wv
❑	Bar-tailed Godwit	*Limosa lapponica*	pm
❑	Whimbrel	*Numenius phaeopus*	pm
❑	Slender-billed Curlew	*Numenius tenuirostris*	v
❑	Eurasian Curlew	*Numenius arquata*	pm, wv
❑	Spotted Redshank	*Tringa erythropus*	pm, wv
❑	Common Redshank	*Tringa totanus*	WV, PM
❑	Marsh Sandpiper	*Tringa stagnatilis*	PM
❑	Common Greenshank	*Tringa nebularia*	PM, wv
❑	Green Sandpiper	*Tringa ochropus*	PM, wv
❑	Wood Sandpiper	*Tringa glareola*	PM
❑	Terek Sandpiper	*Xenus cinereus*	v
❑	Common Sandpiper	*Actitis hypoleucos*	PM, WV
❑	Ruddy Turnstone	*Arenaria interpres*	pm, wv
❑	Red Knot	*Calidris canutus*	pm
❑	Sanderling	*Calidris alba*	pm, wv
❑	Little Stint	*Calidris minuta*	PM, WV
❑	Temminck's Stint	*Calidris temminckii*	PM, wv
❑	Baird's Sandpiper	*Calidris bairdii*	v
❑	Pectoral Sandpiper	*Calidris melanotos*	v
❑	Curlew Sandpiper	*Calidris ferruginea*	PM
❑	Dunlin	*Calidris alpina*	PM, WV
❑	Broad-billed Sandpiper	*Calidris falcinellus*	pm
❑	Ruff	*Calidris pugnax*	PM, wv

❑	Red-necked Phalarope	*Phalaropus lobatus*	pm
❑	Red Phalarope	*Phalaropus fulicarius*	v
❑	Cream-coloured Courser	*Cursorius cursor*	pm
❑	Collared Pratincole	*Glareola pratincola*	PM, ob
❑	Oriental Pratincole	*Glareola maldivarum*	v
❑	Black-winged Pratincole	*Glareola nordmanni*	pm
❑	Black-legged Kittiwake	*Rissa tridactyla*	v
❑	Slender-billed Gull	*Chroicocephalus genei*	PM, wv
❑	Black-headed Gull	*Chroicocephalus ridibundus*	WV, PM
❑	Little Gull	*Hydrocoloeus minutus*	pm, wv
❑	Franklin's Gull	*Leucophaeus pipixcan*	v
❑	Audouin's Gull	*Ichthyaetus audouinii*	rb/mb?, pm
❑	Mediterranean Gull	*Ichthyaetus melanocephalus*	pm, wv
❑	Pallas's Gull	*Ichthyaetus ichthyaetus*	pm, wv
❑	White-eyed Gull	*Ichthyaetus leucophthalmus*	v
❑	Common Gull	*Larus canus*	wv
❑	Great Black-backed Gull	*Larus marinus*	v
❑	European Herring Gull	*Larus argentatus*	v
❑	Caspian Gull	*Larus cachinnans*	WV, pm
❑	Yellow-legged Gull	*Larus michahellis*	RB, PM, WV
❑	Armenian Gull	*Larus armenicus*	WV, pm
❑	Lesser Black-backed Gull	*Larus fuscus*	PM, wv
❑	– Baltic Gull	*Larus fuscus fuscus*	PM, wv
❑	– Heuglin's Gull	*Larus fuscus heuglini*	pm, wv
❑	Gull-billed Tern	*Gelochelidon nilotica*	pm
❑	Caspian Tern	*Hydroprogne caspia*	pm
❑	Sandwich Tern	*Thalasseus sandvicensis*	PM, WV
❑	Little Tern	*Sternula albifrons*	pm, ob
❑	Common Tern	*Sterna hirundo*	pm, ob
❑	Arctic Tern	*Sterna paradisaea*	v
❑	Whiskered Tern	*Chlidonias hybrida*	PM
❑	White-winged Tern	*Chlidonias leucopterus*	PM
❑	Black Tern	*Chlidonias niger*	pm
❑	Pomarine Skua	*Stercorarius pomarinus*	v
❑	Arctic Skua	*Stercorarius parasiticus*	pm
❑	Pin-tailed Sandgrouse	*Pterocles alchata*	v
❑	Black-bellied Sandgrouse	*Pterocles orientalis*	wv, FB
❑	Rock Dove	*Columba livia*	RB
❑	Stock Dove	*Columba oenas*	wv, pm

❑	Common Wood Pigeon	*Columba palumbus*	RB, WV
❑	European Turtle Dove	*Streptopelia turtur*	PM, MB
❑	Oriental Turtle Dove	*Streptopelia orientalis*	v
❑	Eurasian Collared Dove	*Streptopelia decaocto*	RB, pm?
❑	Laughing Dove	*Spilopelia senegalensis*	rb, v, E/I?
❑	Namaqua Dove	*Oena capensis*	v
❑	Great Spotted Cuckoo	*Clamator glandarius*	PM, MB
❑	Diederik Cuckoo	*Chrysococcyx caprius*	v
❑	Common Cuckoo	*Cuculus canorus*	PM, mb
❑	Western Barn Owl	*Tyto alba*	RB
❑	Eurasian Scops Owl	*Otus scops*	pm, wv?
❑	Cyprus Scops Owl	*Otus cyprius*	RB, mb?
❑	Little Owl	*Athene noctua*	RB
❑	Long-eared Owl	*Asio otus*	RB, pm, wv?
❑	Short-eared Owl	*Asio flammeus*	pm, wv
❑	European Nightjar	*Caprimulgus europaeus*	PM, MB
❑	Alpine Swift	*Tachymarptis melba*	PM, MB
❑	Common Swift	*Apus apus*	PM, MB
❑	Pallid Swift	*Apus pallidus*	PM, MB
❑	Little Swift	*Apus affinis*	pm, ob?
❑	European Roller	*Coracias garrulus*	PM, MB
❑	White-throated Kingfisher	*Halcyon smyrnensis*	v
❑	Common Kingfisher	*Alcedo atthis*	PM, WV
❑	Pied Kingfisher	*Ceryle rudis*	pm, wv, ob
❑	Blue-cheeked Bee-eater	*Merops persicus*	pm
❑	European Bee-eater	*Merops apiaster*	PM, mb
❑	Eurasian Hoopoe	*Upupa epops*	PM, MB
❑	Eurasian Wryneck	*Jynx torquilla*	PM, wv
❑	Lesser Kestrel	*Falco naumanni*	PM, FB
❑	Common Kestrel	*Falco tinnunculus*	RB, PM, WV?
❑	Red-footed Falcon	*Falco vespertinus*	PM
❑	Amur Falcon	*Falco amurensis*	v
❑	Eleonora's Falcon	*Falco eleonorae*	MB, PM
❑	Sooty Falcon	*Falco concolor*	v
❑	Merlin	*Falco columbarius*	pm, wv
❑	Eurasian Hobby	*Falco subbuteo*	PM, mb
❑	Lanner Falcon	*Falco biarmicus*	v
❑	Saker Falcon	*Falco cherrug*	pm
❑	Peregrine Falcon	*Falco peregrinus*	RB, PM, WV

❑	Red-backed Shrike	*Lanius collurio*	PM, ob
❑	Isabelline Shrike	*Lanius isabellinus*	pm
❑	Red-tailed Shrike	*Lanius phoenicuroides*	pm
❑	Lesser Grey Shrike	*Lanius minor*	PM
❑	Great Grey Shrike	*Lanius excubitor*	v
❑	– Saharan Grey Shrike	*Lanius excubitor elegans*	v
❑	– Arabian Grey Shrike	*Lanius excubitor aucheri*	v
❑	Steppe Grey Shrike	*Lanius pallidirostris*	v
❑	Woodchat Shrike	*Lanius senator*	PM, mb
❑	Masked Shrike	*Lanius nubicus*	PM, MB
❑	Eurasian Golden Oriole	*Oriolus oriolus*	PM, mb
❑	Eurasian Jay	*Garrulus glandarius*	RB
❑	Eurasian Magpie	*Pica pica*	RB, wv?
❑	Yellow-billed Chough	*Pyrrhocorax graculus*	v
❑	Western Jackdaw	*Corvus monedula*	RB, wv?
❑	House Crow	*Corvus splendens*	v
❑	Rook	*Corvus frugilegus*	v
❑	Hooded Crow	*Corvus cornix*	RB
❑	Brown-necked Raven	*Corvus ruficollis*	v
❑	Northern Raven	*Corvus corax*	rb
❑	Bohemian Waxwing	*Bombycilla garrulus*	v
❑	Hypocolius	*Hypocolius ampelinus*	v
❑	Coal Tit	*Periparus ater*	RB, v
❑	Great Tit	*Parus major*	RB
❑	Eurasian Penduline Tit	*Remiz pendulinus*	PM, wv, FB?
❑	Bearded Reedling	*Panurus biarmicus*	pm, wv
❑	Calandra Lark	*Melanocorypha calandra*	RB, pm, wv
❑	Bimaculated Lark	*Melanocorypha bimaculata*	pm
❑	Bar-tailed Lark	*Ammomanes cinctura*	v
❑	Greater Short-toed Lark	*Calandrella brachydactyla*	PM, ob
❑	Lesser Short-toed Lark	*Alaudala rufescens*	pm, wv
❑	(African) Dunn's Lark	*Eremalauda dunni*	v
❑	Crested Lark	*Galerida cristata*	RB
❑	Woodlark	*Lullula arborea*	WV, pm, rb
❑	Eurasian Skylark	*Alauda arvensis*	WV, PM, ob
❑	Horned Lark	*Eremophila alpestris*	v
❑	Temminck's Lark	*Eremophila bilopha*	v
❑	Sand Martin	*Riparia riparia*	PM
❑	Barn Swallow	*Hirundo rustica*	PM, MB

❑	Eurasian Crag Martin	*Ptyonoprogne rupestris*	rb, pm, wv
❑	Common House Martin	*Delichon urbicum*	MB, PM
❑	Red-rumped Swallow	*Cecropis daurica*	PM, MB
❑	Cetti's Warbler	*Cettia cetti*	RB
❑	Willow Warbler	*Phylloscopus trochilus*	PM
❑	Common Chiffchaff	*Phylloscopus collybita*	PM, WV
❑	Mountain Chiffchaff	*Phylloscopus sindianus*	v
❑	Eastern Bonelli's Warbler	*Phylloscopus orientalis*	PM
❑	Wood Warbler	*Phylloscopus sibilatrix*	PM
❑	Dusky Warbler	*Phylloscopus fuscatus*	v
❑	Yellow-browed Warbler	*Phylloscopus inornatus*	v
❑	Green Warbler	*Phylloscopus nitidus*	v
❑	Basra Reed Warbler	*Acrocephalus griseldis*	v
❑	Great Reed Warbler	*Acrocephalus arundinaceus*	PM, ob
❑	Moustached Warbler	*Acrocephalus melanopogon*	wv
❑	Sedge Warbler	*Acrocephalus schoenobaenus*	PM
❑	Paddyfield Warbler	*Acrocephalus agricola*	v
❑	Blyth's Reed Warbler	*Acrocephalus dumetorum*	v
❑	Eurasian Reed Warbler	*Acrocephalus scirpaceus*	MB, PM
❑	Marsh Warbler	*Acrocephalus palustris*	pm
❑	Eastern Olivaceous Warbler	*Iduna pallida*	MB, PM
❑	Upcher's Warbler	*Hippolais languida*	v
❑	Olive-tree Warbler	*Hippolais olivetorum*	pm
❑	Icterine Warbler	*Hippolais icterina*	pm
❑	Common Grasshopper Warbler	*Locustella naevia*	v
❑	River Warbler	*Locustella fluviatilis*	v
❑	Savi's Warbler	*Locustella luscinioides*	pm
❑	Zitting Cisticola	*Cisticola juncidis*	RB
❑	Eurasian Blackcap	*Sylvia atricapilla*	PM, WV, ob?
❑	Garden Warbler	*Sylvia borin*	PM
❑	Barred Warbler	*Sylvia nisoria*	pm
❑	Lesser Whitethroat	*Sylvia curruca*	PM
❑	Eastern Orphean Warbler	*Sylvia crassirostris*	PM
❑	Asian Desert Warbler	*Sylvia nana*	v
❑	Common Whitethroat	*Sylvia communis*	PM
❑	Spectacled Warbler	*Sylvia conspicillata*	RB, mb?
❑	Subalpine Warbler	*Sylvia cantillans*	PM
❑	Sardinian Warbler	*Sylvia melanocephala*	RB, PM, WV
❑	Ménétries's Warbler	*Sylvia mystacea*	v

❑	Rüppell's Warbler	*Sylvia rueppelli*	PM
❑	Cyprus Warbler	*Sylvia melanothorax*	RB, MB
❑	Firecrest	*Regulus ignicapilla*	v
❑	Goldcrest	*Regulus regulus*	pm, wv
❑	Eurasian Wren	*Troglodytes troglodytes*	RB, wv?
❑	Wallcreeper	*Tichodroma muraria*	wv, pm?
❑	Short-toed Treecreeper	*Certhia brachydactyla*	RB
❑	Rose-coloured Starling	*Pastor roseus*	pm
❑	Common Starling	*Sturnus vulgaris*	WV, PM
❑	Ring Ouzel	*Turdus torquatus*	v
❑	Eurasian Blackbird	*Turdus merula*	RB, PM, WV
❑	Naumann's Thrush	*Turdus naumanni*	v
❑	Dusky Thrush	*Turdus eunomus*	v
❑	Fieldfare	*Turdus pilaris*	pm, wv
❑	Redwing	*Turdus iliacus*	wv, pm
❑	Song Thrush	*Turdus philomelos*	WV, PM
❑	Mistle Thrush	*Turdus viscivorus*	wv, pm
❑	Rufous-tailed Scrub Robin	*Cercotrichas galactotes*	pm
❑	Spotted Flycatcher	*Muscicapa striata*	PM, mb
❑	European Robin	*Erithacus rubecula*	PM, WV
❑	Bluethroat	*Luscinia svecica*	PM, WV
❑	Thrush Nightingale	*Luscinia luscinia*	pm
❑	Common Nightingale	*Luscinia megarhynchos*	PM, MB
❑	White-throated Robin	*Irania gutturalis*	v
❑	Red-flanked Bluetail	*Tarsiger cyanurus*	v
❑	Eurasian Pied Flycatcher	*Ficedula hypoleuca*	PM
❑	Collared Flycatcher	*Ficedula albicollis*	PM
❑	Semi-collared Flycatcher	*Ficedula semitorquata*	pm
❑	Red-breasted Flycatcher	*Ficedula parva*	pm
❑	Black Redstart	*Phoenicurus ochruros*	WV, PM
❑	– Central Asian Black Redstart	*Phoenicurus ochruros phoenicuroides*	v
❑	Common Redstart	*Phoenicurus phoenicurus*	PM
❑	– Ehrenberg's Redstart	*Phoenicurus phoenicurus samamisicus*	pm
❑	Blue Rock Thrush	*Monticola solitarius*	PM, WV, rb
❑	Common Rock Thrush	*Monticola saxatilis*	pm
❑	Whinchat	*Saxicola rubetra*	PM
❑	European Stonechat	*Saxicola rubicola*	WV, PM
❑	Siberian Stonechat	*Saxicola maurus*	pm

❑	– Caspian Stonechat	*Saxicola maurus hemprichii*	pm
❑	Pied Bush Chat	*Saxicola caprata*	v
❑	Northern Wheatear	*Oenanthe oenanthe*	PM
❑	Isabelline Wheatear	*Oenanthe isabellina*	PM
❑	Hooded Wheatear	*Oenanthe monacha*	pm
❑	Desert Wheatear	*Oenanthe deserti*	pm
❑	Black-eared Wheatear	*Oenanthe hispanica*	PM
❑	Cyprus Wheatear	*Oenanthe cypriaca*	MB
❑	Pied Wheatear	*Oenanthe pleschanka*	v
❑	White-crowned Wheatear	*Oenanthe leucopyga*	v
❑	Finsch's Wheatear	*Oenanthe finschii*	WV, pm
❑	Mourning Wheatear	*Oenanthe lugens*	v
❑	Kurdish Wheatear	*Oenanthe xanthoprymna*	v
❑	White-throated Dipper	*Cinclus cinclus*	FB
❑	House Sparrow	*Passer domesticus*	RB, PM? wv?
❑	Spanish Sparrow	*Passer hispaniolensis*	RB, PM, WV
❑	Dead Sea Sparrow	*Passer moabiticus*	FB, v
❑	Eurasian Tree Sparrow	*Passer montanus*	v
❑	Pale Rockfinch	*Carpospiza brachydactyla*	v
❑	Rock Sparrow	*Petronia petronia*	v, wv?
❑	White-winged Snowfinch	*Montifringilla nivalis*	v
❑	Dunnock	*Prunella modularis*	wv, pm?
❑	Western Yellow Wagtail	*Motacilla flava*	PM, mb
❑	– Yellow-crowned (British) Yellow Wagtail	*Motacilla flava flavissima*	v
❑	– Yellow-headed Wagtail	*Motacilla flava lutea*	v
❑	– Blue-headed Wagtail	*Motacilla flava flava*	PM
❑	– Sykes's Wagtail	*Motacilla flava beema*	pm
❑	– Ashy-headed Wagtail	*Motacilla flava cinereocapilla*	v
❑	– Black-headed Wagtail	*Motacilla flava feldegg*	PM, mb
❑	– Eastern Black-headed (White-chinned) Wagtail	*Motacilla flava melanogrisea*	v
❑	– Grey-headed Wagtail	*Motacilla flava thunbergi*	PM
❑	Citrine Wagtail	*Motacilla citreola*	pm
❑	Grey Wagtail	*Motacilla cinerea*	PM, WV, rb
❑	White Wagtail	*Motacilla alba*	PM, WV, ob
❑	– Pied Wagtail	*Motacilla alba yarrellii*	v
❑	– Masked Wagtail	*Motacilla alba personata*	v
❑	Richard's Pipit	*Anthus richardi*	pm
❑	Blyth's Pipit	*Anthus godlewskii*	v

❑	Tawny Pipit	*Anthus campestris*	PM
❑	Long-billed Pipit	*Anthus similis*	v
❑	Meadow Pipit	*Anthus pratensis*	WV, PM
❑	Tree Pipit	*Anthus trivialis*	PM
❑	Olive-backed Pipit	*Anthus hodgsoni*	v
❑	Red-throated Pipit	*Anthus cervinus*	PM, wv
❑	Buff-bellied Pipit	*Anthus rubescens*	v
❑	Water Pipit	*Anthus spinoletta*	PM, WV
❑	Common Chaffinch	*Fringilla coelebs*	RB, PM, WV
❑	Brambling	*Fringilla montifringilla*	wv, pm
❑	Hawfinch	*Coccothraustes coccothraustes*	wv
❑	Eurasian Bullfinch	*Pyrrhula pyrrhula*	v
❑	Eurasian Crimson-winged Finch	*Rhodopechys sanguineus*	v
❑	Trumpeter Finch	*Bucanetes githagineus*	pm
❑	Common Rosefinch	*Carpodacus erythrinus*	v
❑	European Greenfinch	*Carduelis chloris*	RB, PM, WV
❑	Desert Finch	*Rhodospiza obsoleta*	v
❑	Common Linnet	*Linaria cannabina*	RB, WV, PM
❑	Common Redpoll	*Acanthis flammea*	v
❑	Red Crossbill	*Loxia curvirostra*	RB, wv?
❑	European Goldfinch	*Carduelis carduelis*	RB, WV, PM
❑	Red-fronted Serin	*Serinus pusillus*	v
❑	European Serin	*Serinus serinus*	PM, WV, RB
❑	Eurasian Siskin	*Spinus spinus*	wv
❑	Corn Bunting	*Emberiza calandra*	RB, PM, WV
❑	Yellowhammer	*Emberiza citrinella*	wv
❑	Pine Bunting	*Emberiza leucocephalos*	v
❑	Rock Bunting	*Emberiza cia*	v, pm?, wv?
❑	Cinereous Bunting	*Emberiza cineracea*	pm
❑	Ortolan Bunting	*Emberiza hortulana*	PM
❑	Yellow-breasted Bunting	*Emberiza aureola*	v
❑	Cretzschmar's Bunting	*Emberiza caesia*	MB, PM
❑	Black-headed Bunting	*Emberiza melanocephala*	MB, PM
❑	Common Reed Bunting	*Emberiza schoeniclus*	WV, pm
❑	Little Bunting	*Emberiza pusilla*	v
❑	Rustic Bunting	*Emberiza rustica*	v

INDEX

C

G

H

I

J

K

L

M

N

O

P

Q

R

T

U

V